The Ecology of Industry

Sectors and Linkages

Edited by

Deanna J. Richards and Greg Pearson

NATIONAL ACADEMY OF ENGINEERING

NATIONAL ACADEMY PRESS
Washington, D.C. 1998

NATIONAL ACADEMY PRESS • 2101 Constitution Avenue, N.W. • Washington, DC 20418

The National Academy of Engineering was established in 1964, under the charter of the National Academy of Sciences, as a parallel organization of outstanding engineers. It is autonomous in its administration and in the selection of its members, sharing with the National Academy of Sciences the responsibility for advising the federal government. The National Academy of Engineering also sponsors engineering programs aimed at meeting national needs, encourages education and research, and recognizes the superior achievements of engineers. Wm. A. Wulf is president of the National Academy of Engineering.

This volume has been reviewed by a group other than the authors according to procedures approved by a National Academy of Engineering report review process. The interpretations and conclusions expressed in the papers are those of the authors and are not presented as the views of the council, officers, or staff of the National Academy of Engineering.

Funding for the activity that led to this publication was provided by the Ralph M. Parsons Foundation, the AT&T Foundation, the Andrew W. Mellon Foundation, and the National Academy of Engineering Technology Agenda Program.

Library of Congress Cataloging Card Number: 98-84941
International Standard Book Number: 0-309-06355-8

Cover art: *Distorted Eye,* courtesy of the artist, Grace Selvanayagam, Kuala Lumpur, Malaysia.

Copyright 1998 by the National Academy of Sciences. All rights reserved.

Printed in the United States of America

Steering Committee

ROBERT A. FROSCH (*Chairman*), Senior Research Fellow, Harvard University (Former Vice President for Research, General Motors)
PETER R. BRIDENBAUGH, Executive Vice President, Automotive Aluminum Company of America
ROBERT C. FORNEY, Retired Executive Vice President, E.I. du Pont de Nemours & Company
G. FRANK JOKLIK, Retired President and CEO, Kennecott Corporation
ROBERT A. LAUDISE, Director, Materials and Processing Research Laboratory, Lucent Technology
LEE THOMAS, Senior Vice President, Environmental Government Affairs, Georgia-Pacific Corporation
KURT YEAGER, President, Electric Power Research Institute

Staff

DEANNA J. RICHARDS, Project Director
GREG PEARSON, Editor
MARION RAMSEY, Senior Program Assistant
LONG NGUYEN, Project Assistant

Preface

Technology has played and will continue to play an important role in economic development and environmental protection. The industries featured in this volume have been critical to economic development and, together with other industries, represent a mosaic of economic activity that defines the ecology of industry. As humanity works toward sustainable development—and addresses economic, environmental, and social concerns in an integrated manner—solutions will be increasingly dependent on technological advances made possible by engineering. The application of technology in the service of environmental goals by the industries featured in this volume attests to that fact.

Deanna J. Richards and Robert A. Frosch, in the Overview and Perspectives chapter, provide an even broader view of the role of technological innovation in environmental protection, and they speculate on the role the information revolution will have in shaping the ecology of industry well into the future. Technological advances, however, do not occur in a vacuum. The application of technology to specific situations in industry and the development of technologies that from time to time alter the technological and societal landscape result from long-term investments by industry, government, and the public. As long as that support continues, technological advances will provide the backbone for a sustainable future.

The papers in this report were developed in several sectoral working sessions organized as part of an international conference on industrial ecology convened by the National Academy of Engineering (NAE) in May 1994. Collectively, they provide industrial perspectives on the state of practice and opportunities for improvement in designing and managing for the environment. The overview was developed based on those papers and subsequent workshops held by NAE in

1995 on the impact of services industries and the environment, in 1996 on polymers and the environment, and in 1997 on information technology and the environment.

Many individuals were involved in the preparation of this volume. On behalf of NAE, I want to thank particularly the authors for their thoughtful contributions and the members of the conference steering committee—Peter R. Bridenbaugh, Robert Forney, Robert A. Frosch (chairman), Frank Joklik, Robert Laudise, Lee Thomas, and Kurt Yeager—for their help in organizing the conference.

I would also like to thank the NAE staff who worked on this project. Deanna J. Richards, associate director of the NAE Program Office, heads NAE's Technology and Environment effort and was primarily responsible for shepparding the project through its various stages. Thanks also go to the editorial team that worked on this volume: Greg Pearson, the Academy's editor, contributed invaluable and steadfast editing and publishing oversight with the able assitance of Marion Ramsey, senior program assistant, who provided both editorial support and critical logistical and administrative support, and Long Nguyen, project assistant, who finalized the document for publication. Thanks also go to Bruce Guile, former director of the NAE Program Office, for his contribution to the genesis and oversight of this project during his tenure.

Finally, I would like to express my appreciation to the AT&T Foundation and the Ralph M. Parsons Foundation for their partial support of this project and to the Andrew W. Mellon Foundation for its partial support of related elements of the Academy's Technology and Environment Program.

> WM. A. WULF
> President
> National Academy of Engineering

Acknowledgments

The industry papers in this volume reflect material presented and discussed at sectoral working groups organized at the National Academy of Engineering (NAE) International Conference on Industrial Ecology, May 9–12, 1994 in Irvine, California. The papers were subsequently revised by the authors. The NAE gratefully acknowledges the individuals who participated in the working groups and the authors of the industry papers for their patience and tireless efforts.

THE EXTRACTIVE INDUSTRIES

G. Frank Joklik, Kennecott Corp. (retired) *(Chair)*; Ray Beebe, consultant; Preston Chiaro, Kennecott Corp.; David Cobb, Bechtel Corp.; Carmen Gonzalez, Mexican Petroleum Institute; Henning Grann, Statoil Development Project, Germany; John Innes, CRA Limited, Australia; Sue Kiser, WZI; Eduardo Loreto, Faculty of Engineering of the National University of Mexico; Hiroshi Sakurai, The Engineering Academy of Japan; Peter Schulze, Austin College *(Rapporteur)*; and Mary Jane Wilson, WZI.

PRIMARY MATERIALS PROCESSING

Peter Bridenbaugh, Aluminum Company of America *(Chair)*; Braden Allenby, AT&T; Frederick Anderson, Cadwalader, Wickersham & Taft; Yumi Akimoto, Mitsubishi Material Corp., Japan; Patrick Atkins, Aluminum Company of America; Andre Cabelli, Commonwealth Scientific and Industrial Research Organization, Australia; David Cahn, California Portland Cement Company; Craig Campbell, American Portland Cement Alliance; Gordon Forward, Chaparral Steel Company; Ann Fullerton, The Fullerton Group *(Rapporteur)*; Corina Hebestreit, European Aluminum Association; Rolf Marstrander, Hydro Aluminum a.s., Norway; Kenneth Martchek, Aluminum Company of America; Tomomi Murata, Nippon Steel Corp., Japan; Donald Sadoway, Massachusetts Institute of Technology; and Roger Strelow, Bechtel Corp.

MANUFACTURING

Robert Laudise, AT&T *(Chair)*; Jan Agri, the Royal Swedish Academy of Engineering Sciences; Patrick Eagan, Engineering Professional Development; Wojciech Gaweda, Academy of Engineering of Poland; Sukehiro Gotoh, National Institute for Environmental Studies, Japan; Thomas Graedel, Yale University; Janos Hadas, Hungarian Academy of Engineering; Joseph Heim, University of Washington at Seattle *(Rapporteur)*; Lisbeth Valentin Hansen, Danish Toxicology Center; Sverker Hogberg, Swedish Waste Research Council; Inge Horkeby, AB Volvo, Sweden; Kosuke Ishii, The Ohio State University; Gregory Keoleian, University of Michigan; Kurt Lange, University of Stuttgart, Germany; David Marks, Massachusetts Institute of Technology; William Moore, Dames and Moore; Steven Moran, Alberta Research Council, Canada; M. Granger Morgan, Carnegie Mellon University; Peter Peterson, U.S. Steel Group; Christine Rosen, University of California; Paul Sheng, University of California; Ernest Smerdon, University of Arizona; Walter Stahel, the Product-Life Institute, Switzerland; Brian Steelman, Ciba-Geigy Corp.; John Stemniski, Charles Stark Draper Laboratory; Masazumi Tanazawa, Toyota Motor Corp., Japan; and Bruce Vigon, Battelle National Laboratory.

ELECTRIC UNTILITIES

Ian Torrens, Electric Power Research Institute *(Chair)*; Jesse Ausubel, Rockefeller University; Alberto Bustani, Instituto Tecnológico y de Estudios Superiores de Monterrey, Mexico; David Cope, the UK Centre for Economic and Environmental Development; Glyn England, Windcluster Limited, United Kingdom; Jørgen Kjems, Forskningscenter Risø, Denmark; Richard Macve, University of Wales; Kenji Matsuo, Tokyo Electric Power Company; William Moore, Dames and Moore; Yasuo Mori, Tokyo Institute of Technology; Nobuyuki Nishikawa, Tokyo Electric Power Company; David Rejeski, U.S. Environmental Protection Agency; Don Roberts, CH2M Hill; Richard Schuler, Cornell University; Deanna Richards, National Academy of Engineering of the United States *(Rapporteur)*.

PULP AND PAPER

W.C. Pete Howard, Georgia-Pacific Corp. *(Chair)*; Doug Armstrong, Georgia-Pacific Corp.; Jens Folke, NfiljoForskerGruppen ApS., Denmark; Isaiah Gellman, National Council for Air and Stream Improvement of the United States; Robert Johnston, Australian Pulp and Paper Institute; John Lee Jr., CH2M Hill; William Nicholson, Potlatch Corp.; and Raymond Wassel, National Research Council of the United States *(Rapporteur)*.

Contents

Preface	v
Overview and Perspectives *Deanna J. Richards and Robert A. Frosch*	1
The Extractive Industries *Preston S. Chiaro and G. Frank Joklik*	13
Primary Materials Processing *Charles G. Carson III, Patrick R. Atkins, Elizabeth H. Mikols, Kenneth J. Martchek, and Ann B. Fullerton*	27
Manufacturing *Robert A. Laudise and Thomas E. Graedel*	45
The Electric Utility Industry *Ian M. Torrens and Kurt E. Yeager*	72
The Pulp and Paper Industry *A. Douglas Armstrong, Keith M. Bentley, Sergio F. Galeano, Robert J. Olszewski, Gail A. Smith, and Jonathan R. Smith Jr.*	101
Biographical Data	142

Overview and Perspectives

DEANNA J. RICHARDS AND ROBERT A. FROSCH

The ecological analog for industry, like any analogy, is not perfect. There are striking similarities and obvious differences between natural and industrial systems.[1] In this overview, the term "ecology of industry" is used descriptively. Hence, industry's ecology is defined by the metabolism of materials (the flow of materials through industrial systems, including their transformations during flow); the use of energy, labor, and capital; and the application of information or knowledge. A characteristic of ecological systems is that they evolve. The evolution of industrial systems and their use (and storage) of resources are affected by the introduction of new technologies, decisions made in design, preferences of consumers, regulatory dictates, and the like. The idea of industry having an ecological structure is not new. The application of this concept, however, is receiving greater scrutiny[2] as, for example, publications have emerged that explore the various dimensions of industrial ecology. Indeed, "industrial ecology" has become jargon for describing systems of production and consumption networks that have a minimal impact on the environment as a primary objective and have an overarching objective of environmentally sustainable economic expansion.

This volume builds on earlier efforts of the National Academy of Engineering (NAE) in the area of technology and the environment.[3] It presents industrial perspectives on opportunities and challenges in improving environmental efficiencies through better design and management in five industries: mining, materials processing, manufacturing, electric utilities, and pulp and paper. The accompanying papers result from sectoral workshops on industrial best practices convened as part of a 1994 NAE International Conference on Industrial Ecology in Irvine, California. This overview also draws on four other NAE workshops: two that examined the impact of the services sector on the environment, held in October 1994 and June 1995; one that examined the impact of synthetic polymers on the environment, held in August 1996; and one that explored the implications of information technology for the environment, held in July 1997.

This overview and perspective considers the stimuli that lead a firm to take environmental action, the role of technology in effecting change in industrial systems, the interconnections between production and consumption activities, and the role of information and knowledge in improving environmental efficiencies.

FIRM MOTIVATION AND EXTERNAL STIMULI

Much progress has been made over the last decade in integrating environmental considerations into the mainstream of business decision making. In the process, the ecology of industry has changed. The impetus for the change can be attributed to four interrelated factors. The first is corporate well-being, which is determined by profits and growth. Hence, it is not unusual to find that the principal factor driving environmental efficiency improvements in industry has been the recognition that inefficiently run production systems result in waste generation and energy losses, which if minimized can lead to cost savings. It was sobering for some companies to realize that the cost of environmental compliance was equal to their entire research and development budget (Carberry, 1997). The rising cost of compliance has provided impetus to companies to develop and implement cleaner technologies that do not emit regulated emissions or that use substitutes for regulated substances to cut the cost of compliance. That approach has also led to the creation of new profit centers in companies that market the cleaner technologies to other firms with similar emissions problems.

The second factor is consumer demand. Demand revolves around the performance and cost of a product or service, as well as the image and social acceptability of a product, service, or company. As consumers place more value on environmental attributes, corporations with cost-cutting environmental strategies are able to market their environmental image in addition to other desirable features of their products or services. Depending on the type of product or service and corporate strategy, consumers can also be involved in the company's environmental efforts. A hotel that encourages its customers to have their towels and sheets changed every other day rather than daily saves on energy and water used for cleaning, while providing an environmental value that customers might seek. In addition, the hotel engages in public education through the decals and notices in rooms about the program and its impacts on water and energy use. Similar benefits result from programs that encourage consumers to return spent printer cartridges to the manufacturer for reprocessing and recycling. In the example of the hotel, there is clear cost savings to the hotel. However, in the case of the printer cartridge, recovery may be driven by product recovery laws, as it is in Germany, and involves additional costs. Some of the additional costs can be avoided if product recovery considerations are incorporated in design and management decisions. Nevertheless, in most cases, although consumers say they want environmentally friendly products, they have not voted with their pocketbooks (Laudise and Graedel, this volume; Simon and Woodell, 1997). This poses a challenge to government and industry: Government needs to encourage the development of more environmentally preferable products; government and industry need to develop and provide the incentives to consumers to buy the products; and industry needs to find ways to improve production and lower product costs.

The third factor influencing the ecology of industry is government. Through regulations and laws, government dictates the bounds of industrial operations. Armstrong et al. (this volume) and Chiaro and Joklik (this volume) point to some of the effects of environmental regulation on the pulp and paper and mining industries, respectively. The rules set by government, however, affect a diverse set of industrial operations: finance (through, for example, the Securities and Exchange Commission and antitrust regulations); labor and workplace practices (through such agencies as the Occupational Safety and Health Administration and the National Labor Relations Board); and consumer protection (through such agencies as the Environmental Protection Agency and the Food and Drug Administration).

Sometimes the rules are in conflict. For example, consumer protection laws that classify products with refurbished parts as secondhand (even if they perform to the same standards as new products) send a wrong signal when it comes to promoting reuse or recycling. Through subsidies, governments also send signals about the importance of certain industrial operations or materials, which would not fare as well in an economy without the subsidies.

In addition, companies with global operations have to increasingly abide by the rules and dictates of many nations. The diversity of environmental rules has given rise to the need for harmonization. An outgrowth of this situation has been the development of a set of environmental standards by the International Standards Organization (ISO). The so-called ISO 14000 standards[4] do not replace regulation. Rather, they are an attempt to establish a set of ground rules for corporate environmental practice.

The final factor influencing the ecology of industry is community acceptance and industrial impact. Communities in which industries operate are more frequently in the news because of potential environmental health concerns. Community acceptance of industrial activity is also influenced by where industry locates its facilities and the impact such siting has on taxes and local economic development, as well as by other sociopolitical factors. In the larger context, the level of community acceptance can result in demand for new government laws and regulations. Indeed, organizations are increasingly involving communities in the planning and design of their plants and operations. Carson et al. (this volume) provide an example of a mining company in Australia that developed a mine restoration plan in response to and in cooperation with local communities.

Directed efforts to improve industrial environmental efficiencies would not be a priority of business without consumer demand, government regulation, and community acceptance, because the costs of materials do not reflect envrionmental impacts. When these forces come into play, however, environmental factors (costs) that were considered external to running a business are internalized, and companies move environmental issues out of the compliance "doghouse" and into strategic focus. In doing so, they change how energy is used and how materials inputs and outputs are managed. In other words, they change the ecology of industry.

THE TECHNOLOGICAL LINK IN THE EVOLUTION OF INDUSTRIAL SYSTEMS

Improvements in environmental efficiencies are a direct result of deploying technology. Historical data (Ausubel and Langford, 1997; Ausubel and Sladovich, 1989) show that advances in technology have contributed to the evolution of systems that are decarbonizing (using less carbon per unit energy produced), dematerializing (using less material per unit product), and increasing in energy efficiency over time. Technology, however, is not a panacea. Much has been written about the unintended consequences of technology (e.g., Tenner, 1996), suggesting the need to understand the impact of technology on systems of production and consumption and to deal with them.

Yet, the positive impact of technology in addressing specific environmental ills is evident in this volume's papers on the pulp and paper and the electric utilities industries. For instance, technological advances in the making of pulp and paper have been instrumental in minimizing pollutants from pulp and paper operations. Electricity generation and distribution are also important to economic growth and to the role of electrotechnologies in addressing environmental concerns.

Most of the technological innovations discussed in this volume have been incremental. They have occurred almost continually, resulting from pressures to reduce costs or meet quality, design, performance, manufacturability, or environmental goals. They are seldom the direct result of any deliberate R&D, although they frequently benefit indirectly from previous R&D conducted for other purposes. Rather, the improvements are outcomes of inventions and improvements suggested by engineers and others directly engaged in the production or service function, or by users and customers.

Freeman (1992) describes three other categories of innovation that contribute to change in industrial systems: radical innovations, technology-system innovations, and techno-economic revolutions. Radical innovations are discontinuous events that result from deliberate R&D. The underlying science and engineering are often incremental, but the deployment of the results in the economic systems leads to radical changes from past production practices. (For example, incremental improvements in canoes did not lead to steamships nor did the developments of glass and paper lead to the creation of plastics.) These innovations are unevenly distributed over industry sectors and over time. However, when they occur, they are the basis for growth of new markets or of significant improvements in the use of inputs (as well as the lower cost and higher quality of the existing product), as in the case of the oxygen steelmaking process.

Technology-system innovations are far-reaching technological changes that affect many branches of the economy. They ultimately create new sectors of economic activity. The clusters of radical innovations that together gave rise to the semiconductor industry are an example of technology-system innovations. Freeman (1992) provides two other examples of clusters of innovations. The first is in synthetic materials and petrochemicals introduced between the 1930s and 1950s. Adjunct developments in machinery for injection molding and extrusion and later innovations in packaging, construction, electrical equipment, agriculture, textiles, clothing, toys, and other applications resulted in a range of interrelated innovations that were not contemplated when the materials and chemicals were first developed. The second example is the cluster of innovations in electrically driven household consumer durables. Here, the availability of electric power and very cheap electric motors combined with new marketing techniques to transform patterns of household expenditures in high-income countries as well as the organization of production in many industries. Both examples are subsets of innovations that literally changed the industrial metabolism of economies, particulary in terms of the flows of material, energy, and capital.

Techno-economic revolutions result in new technology systems that have ubiquitous effects on the economy. These innovations transform production and management throughout the economy. The introduction of electric power (Torrens and Yeager, this volume) is one example. Changes associated with techno-economic revolutions breed clusters of radical and incremental innovations, which might eventually embody new technological systems. The computer and microelectronics are at the heart of such a revolution today.

In considering the ecology of industry, innovations likely to lead to the most pervasive effect on materials or energy savings will probably emerge from technology-system innovations and techno-economic revolutions. Indeed, pervasive changes are already taking place as a result of the current information and communications revolution, which is techno-economic in nature. The impacts of this revolution on the industrial metabolism of the economy and on the ecology of industry are already being felt, particularly in the monitoring and control of emissions and of energy and materials use; in quality and inventory controls; and in the miniaturization of technology.

Many of the energy-saving technologies and the electrotechnologies discussed by Torrens and Yeager (this volume) and most of the recent process changes in the pulp and paper industry discussed by Armstrong et al. (this volume) depend on the incorporation of electronic sensors and monitors, which interact with feedback control systems and small computers. Monitoring and control systems are used in a similar way in fuel systems of automobiles (and other transporation vehicles). Sensing and monitoring instruments are also essential for many regulatory purposes, and instruments are used to detect pollutants.

Information and communications technology also makes possible greater quality and inventory control and helps reduce and eliminate defective or substandard products. This is not a result of the technology itself, but a diffusion of a management philosophy associated with the technology. Reengineering efforts discussed by Carson et al. (this volume) reflect this new managment philosophy, which abhors the wasteful attitudes and traditions of mass production that often tolerated the production of large amounts of scrap, high reject rates, and significant loss of inventory during production. The environmental benefits of applying information technology with new management philosophies extend beyond a single plant to networks of plants, including outsourced activities. However, some of these practices might have negative environmental consequences. For example, just-in-time practices can lead to increased use of transportation (trucks, railroads, and airplanes) and to associated air pollution concerns.

Finally, information and communications technology has also resulted in the miniaturization of technology, leading to less material being used per unit product or function. Compared with old vacuum-tube technology, semiconductor technology is vastly less materials and energy intensive. The value of miniaturization extends beyond the electronics industry. Torrens and Yeager (this volume) point to the potential decentralization of electric power generation resulting from the application of microturbines and fuel cells. On the materials front, there has been a reduction in metal consumption over the last 20 years (Sousa, 1992). The environmental downside of miniaturization is the dissipative nature of materials use. Recovery and separation of complex combinations of materials is more difficult, and many past environmental ills have resulted from the accumulation of materials in the environment over time. Hence, separation technologies might be epected to grow in importance as part of an overall environmental strategy.

How deeply the information and communications revolution will affect environmental performance is unclear. The environmental benefits of these technologies in manufacturing are well known. In transportation, which accounts for up to 26.5 percent of energy use in the United States (U.S. Department of Energy, 1995), the benefits derived from improved logistics and distribution in business applications are significant. Also, it is commonly perceived that telecommuting can reduce the environmental burden associated with commuting to work on a frequent basis. The potential for information and communications technologies to alter transportation patterns, reduce energy use, and improve air quality—even as the technologies are applied to ease congestion—requires a range of government and industry investments and policies to be realized (Transportation Research Board, 1997).

INTERCONNECTED COMPLEX SYSTEMS OF PRODUCTION AND CONSUMPTION

Advances in transportation and communications technology underlie a rapidly increasing globalization of the ecology of industry; the regular flows of commerce—movements of people, raw materials, intermediate goods, final goods, and capital—have become truly transnational. At the same time, technical advances in areas like materials and production technologies are creating new types of companies. Although the world has long had multinational companies, there is no precedent for the large and growing number of companies (or groups of companies) that manage integrated global production systems and aim their products at an increasingly homogenous global market. As a result, supplier chain management and the management of environmental concerns associated with these complex cross-national chains is a cornerstone of management in contemporary business.

Another changing aspect of the ecology of industry is the growth of the services sector, which accounts for 60 percent of U.S. output and employment (U.S. Department of Commerce, 1996). Industries in this sector provide fundamental economic and societal functions such as transportation, banking and finance, health care, public utilities, retail and wholesale trade, education, and entertainment. The technological intensity of service industries (e.g., transportation, telecommunications, and health care services) is often high; the incomes of many who work in service industries are often well above average (e.g., doctors, lawyers, investment bankers, and airline pilots); and there is substantial capital investment by service companies (e.g., transportation firms, communications firms, and national retail chains). In addition, service companies have an enormous impact on the environment. Many have complex production operations that, inefficiently run, can use large amounts of energy and generate huge waste streams. As such, service businesses—like manufacturers, natural resource companies, and agricultural concerns—play a critical role in environmental improvement.

The leverage an industry has on environmental issues depends on its location in the production-consumption continuum. Mining, upstream in most production-consumption activities, has little or no influence on the behavior of customers for their product, who are downstream in materials processing or manufacturing (Chiaro and Joklik, this volume). Yet, the mining industry provides many of the basic ingredients of economic activities. Manufacturers, on the other hand (Laudise and Graedel, this volume), although depending on downstream activities, have enormous influence on their suppliers and others upstream through the procurement of materials and components.

Service businesses have some unique characteristics in terms of their ability to leverage both upstream and downstream. The companies in this industry sector have significant influence on upstream activities through their merchandise purchasing (e.g., WalMart, Kmart, and Target), provision of food service and delivery (e.g., McDonald's), use of logistics and distribution networks (e.g., United States Postal Service, UPS, and FedEx), management of hospitals and hotels (e.g., Marriott), supply of health management services (e.g., HMOs, Baxter Healthcare), and provision of entertainment (e.g., Busch Gardens and Disney theme parks). Because companies in this sector also interact with a large consumer base, they are a source of knowledge about consumer preferences downstream in the production-consumption system and can play a critical role in conveying environmental information to consumers.

Services' upstream leverage on manufacturers and other economic players is evident. As purchasing agents for millions of consumers, these companies exert tremendous influence on their suppliers by creating markets for environmental improvement. Their downstream influence, however, is yet to be tapped fully. Service firms, to be successful, must be very close to their consumers, and several companies in this sector provide their consumers with environmentally relevant information (e.g., Starbucks Coffee, which provides information about its environmental practices; Home Depot, which provides "green" products next to more common brands; some hotels, which provide an environmental explanation with an offer to change hotel sheets or towels less frequently than daily). Firms that provide this sort of consumer education also provide early insights into consumer preferences and regional buying habits. Such businesses create a true market for environmental improvement, with all the effectiveness that implies. Understanding the role of the services sector in the ecology of industry and examining the energy and materials flows through economic systems (to understand the dynamics, efficiencies, and opportunities for environmental improvements) are critical steps for improving environmental and economic efficiencies more broadly.

On a smaller scale, opportunities for improving environmental efficiencies exist among co-located industries and at the regional level. Torrens and Yeager (this volume) describe the city of Kalundborg, Denmark, to show how co-generation (the use of waste heat and water from a power plant by several industries located nearby) and symbiosis (the use of waste materials from one facility as input to another) have created a low-waste production and consumption system. Such arrangements might be vulnerable to breakdown if one of the partners goes out of business or needs to change its processes. Still, the advantages of such arrangements are being tested in several locations in the United States, including Brownsville, Texas, and Chattanooga, Tennessee (President's Council on Sustainable Development, 1994), and these are generating new information about the industrial metabolism of co-located industrial clusters.

The potential benefit of examining the resource flows of any system depends on the availability of methods to describe and analyze the industrial metabolism—flows of energy, material, and capital—at different scales. Setting priorities among actions intended to improve the environmental performance of these systems is more successful if there is an understanding of how the current system operates. Also useful are information and tools that capture environmental value (costs).

THE INFORMATION-KNOWLEDGE CONNECTION

There are several "public goods" challenges that no single company can justify undertaking alone but which can have a dramatic payoff if companies can share costs and responsibilities. The challenges range from articulating technical and management standards that reflect the best strategic environmental approaches and defining criteria for determining environmental impacts and metrics of environmental performance to the potential use and misuse of environmental information. In each area, there are important roles for government, trade associations, and industry, the latter working with universities and environmental public interest groups (individually or in collaborative groups).

With respect to technical and management standards, the papers in this volume review practices in five industries. A codification and endorsement of practices, such as the Responsible Care program of the Chemical Manufacturers Association, might be effective in raising the standards of practice in other industries. ISO 14000 standards might also have the same effect, but their development is occurring in distinct ISO working groups organized around standards for labeling, life-cycle assessments, and environmental performance indicators—three areas that are interrelated and could benefit from an integrated approach. Such an approach would lead to better streamlining of needs in the area of environmental standards of practice.

A persistent and very difficult aspect of standards relates to assessing the environmental impacts of a product or service. This is true even in mining, which has its own set of suppliers of goods and services but does not have the same leverage on a supplier chain as do manufacturers or service providers who are farther up the "value chain," which extends from the mining of raw ore to final product delivery and beyond. In particular, there is little consensus on the criteria for environmental impact assessment within and outside a firm. The development of reliable "green" certification and mechanisms (perhaps a joint government-business effort) that allows standards sharing and cooperation among businesses in obtaining certification could speed the process of providing consumers with environmental information. There is also a serious need for good data, methods, and institutional mechanisms to undertake state-of-the-art, full environmental life-cycle evaluation.

Further definition and application of industrial environmental performance metrics should also aid efforts to improve environmental efficiencies. Performance metrics, together with a definition of business processes, will provide a framework against which audits (perhaps combined financial and environmental audits) ought to be conducted, because they could trigger prompt monitoring, learning, and mitigation activities. It is to be hoped that environmental measures become valued commodities in the same way quality and safety have come to be valued. Often, what is valued is what gets measured. For instance, toxic release inventory (TRI) data led to fundamental changes in the underlying processes from which the TRI data arose. A broader-based set of environmental performance metrics could do the same. There is a need for a consensus on the structure of environmental performance metrics that are useful for management purposes.

CONCLUSION

A fundamental set of changes in the structure of production and consumption in the U.S. and global economies is under way, driven by information, communications, and transportation technologies. This change is evident in the rate of growth of the services industry compared with the growth rate of the industries covered by the papers in this volume. Another change is the increasing importance of environmental considerations in decision making. This will have profound and as yet only dimly perceived economic and environmental implications. In particular, the technologies of transportation, storage, materials handling, and information allow the creation and production of goods and services to be widely dispersed, yet goods themselves can be delivered almost anywhere very rapidly. The production-distribution chain for products and services is changing in ways that are likely to affect large-scale and diverse environmental issues. These include, for instance, energy consumption for transport, effluents from real estate development and redevelopment, and the capacity to select preferable packaging and product materials using life-cycle assessments. Also affected will be the ability to substitute information and control systems for energy and materials and the capacity to extend knowledge about these systems to less-developed countries before they encounter, and burden the planet with, insurmountable environmental problems.

NOTES

1. Although the analogy of industrial systems to natural ecosystems is not perfect, the two systems do share some fundamental features.
   - They are both composed of individual interacting units, each of which is driven to increase its size or number.
   - The actors in both cases depend upon supplies of resources—in particular, energy, materials, and means of removing or decontaminating by-products.
   - The interactions of the individual actors result in complex patterns of energy and materials flows.

   They also have important differences.
   - Materials cycles are essentially closed in mature natural systems but not in industrial systems.
   - The total productivity of mature natural systems is essentially constant, whereas the productivity of industrial systems tends to grow exponentially.
   - Mature natural systems are more or less sustainable, whereas industrial systems appear nonsustainable as currently configured, either because of production and consumpion patterns or because of their impacts on the environment.
   - The pace of change in mature natural systems is relatively slow—on an evolutionary time scale—but the technology of industrial systems changes rapidly; these changes affect the systems themselves as well as the relationship between industry and the environment.

- Interactions in industrial systems are often mediated by long-range transport, whereas interactions in natural systems generally occur between organisms in proximity to each other.
- The decomposers of natural systems are small and ubiquitous, whereas the decomposers (e.g., recyclers and waste disposers) of industrial systems are relatively few and far between.
- The information content, flows, and feedback loops (e.g., DNA and self-regulation through chemical signals) in natural systems are much more complex and richer than those of industrial or manmade systems.

2. O'Rourke et al. (1996) provide a useful critique of industrial ecology.
3. Over the last several years, NAE has, via its program on Technology and Environment, explored technology's impact on the environment through its role in production and consumption. Recent efforts in this area have focused on technology transfer (*Cross-Border Technology Transfer to Eliminate Ozone-Depleting Substances,* 1992); the effect of science and environmental regulation on technological innovation (*Keeping Pace with Science and Engineering: Case Studies in Environmental Regulation,* 1993); industrial ecology and design for the environment (*The Greening of Industrial Ecosystems,* 1994); corporate environmental practices (*Corporate Environmental Practices: Climbing the Learning Curve,* 1994); the United States' and Japan's interest in industrial ecology (*Industrial Ecology: U.S./Japan Perspectives,* 1994); linkages between natural ecosystem conditions and engineering (*Engineering Within Ecological Constraints,* 1996); design and management of production and consumption systems for environmental quality (*The Industrial Green Game: Implications for Environmental Design and Management,* 1997); an examination of industrial performance measures and their relation to ecosystem conditions (*Measures of Environmental Performance and Ecosystem Condition,* forthcoming 1998); the diffusion patterns of environmentally critical technologies and their effect on the changing habitability of the planet (*Technological Trajectories and the Human Environment,* 1997); the impact of services industries on the environment (exploratory workshops held in October 1994 and June 1995); the impact of polymers on the environment (exploratory workshop held in September 1996); and the role of information and knowledge management in improving environmental efficiencies (workshop held in July 1997).
4. The initial standards foreseen in the ISO 14000 series are:

   ISO 14001: Environmental Management Systems (EMS)—specification with guidance for use.
   ISO 14004: Environmental Management Systems—general guidelines on principles, systems, and supporting techniques.
   ISO 14010-12: Principles, qualification criteria, and procedures for internal and external auditing.
   ISO 14014: Initial review guideline to determine a corporation's baseline operating position, typically used prior to establishing an EMS.
   ISO 14031: Guidance for measuring environmental performance over time.
   ISO 14020-22: Description of labeling principles such as self-declarations of environmental benefits of products.
   ISO 14040-43: Establishment of a methodology for a product's life cycle, including an assessment of impacts and an improvement analysis.
   ISO 14050: Terms and Definitions—ISO standards summary

REFERENCES

Ausubel, J.H., and H.D. Langford, eds. 1997. Technological Trajectories and the Human Environment. Washington, D.C.: National Academy Press.
Ausubel, J.H., and H.E. Sladovich, eds. 1989. Technology and Environment. Washington, D.C.: National Academy Press.

Carberry, J. 1997. Using environmental knowledge systems at DuPont. Paper presented at the 1997 NAE Industrial Ecology Workshop, July 20–22, 1997. Woods Hole, Mass.

Freeman, C. 1992. The Economics of Hope. Essays on Technical Change, Economic Growth, and the Environment. London: Pinter Publications, Ltd.

O'Rouke, D., L. Connelly, and C.P. Koshland. 1996. Industrial ecology. A critical review. International Journal on Environment and Pollution 6(2/3):89–112.

President's Council on Sustainable Development (PCSD). 1994. Eco-Efficiency Task Force Report. Washington, D.C.: PCSD

Simon, F., and M. Woodell. 1997. Consumer perceptions of environmentalism in the triad. Pp. 212–224 in The Industrial Green Game: Implications for Environmental Design and Management, D.J. Richards, ed. Washington, D.C.: National Academy Press.

Sousa, L. J. 1992. Toward a new material paradigm. Minerals Issues (December).

Tenner, E. 1996. Why Things Bite Back: Technology and the Revenge of Unintended Conseqeunces. New York: Knopf.

Transportation Research Board. 1997. Toward a Sustainable Future: Addressing the Long-Term Effects of Motor Vehicle Transportation on Climate and Ecology. National Research Council. Washington, D.C.: National Academy Press.

U.S. Department of Commerce (DOC). 1996. Service Industries and Economic Performance. Washington, D.C.: DOC.

U.S. Department of Energy (DOE). 1995. Annual Energy Review. Washington, D.C.: DOE.

The Extractive Industries

PRESTON S. CHIARO AND G. FRANK JOKLIK

INTRODUCTION

Extractive industries are commonly viewed as having unacceptable impacts on the environment. By their very nature, these industries use energy and disturb the land in extracting the resource being developed. Sustainable development of an extracted resource is a paradoxical concept. Further, there appears to be an inherent, economically based conflict between the extraction of virgin materials and the reduction in the amount of use, reuse, or recycling of these same materials. Indeed, reduction, reuse, and recycling can be viewed as competitors to the extractive industries. How are these apparent conflicts reconciled?

Industrial ecology takes a systems view of the connections between industries. This view embraces all inputs and outputs of energy and materials. One way to reconcile the apparent conflicts described above is to view the extractive industries as an isolated system. The life cycle of such a system is then limited to the material in question but does not extend to any products derived from it. Then an attempt can be made to minimize the amount of energy and resources that go into extracting a particular raw material and to minimize the amount of waste that is created. To address the issue of sustainable development, defined by the World Commission on Environment and Development (1987) as "development that meets the needs of the present without compromising the ability of future generations to meet their own needs," extractive industries, by locating new resources and developing more efficient means of extracting and processing the raw materials, enable future generations to enjoy the benefits of these resources. Further, there are multiple possible uses for the land from which the raw materials are extracted. Taking care in the extraction, processing, and transportation of the raw materials can maximize alternative-use options for the lands from which the materials were extracted. The objective should be to extract and process materials within environmental constraints and to maintain or successfully restore the other values of the site after the resource has been extracted.

This paper raises some key industrial-ecology-related issues of the extractive industries, particularly mining. Historical, current, and potential future practices are examined in an attempt to identify opportunities for and barriers to improvement. This is not an exhaustive review. Rather, the focus is on four topics: environmental stewardship; environmental regulations; life-cycle practices; and cultural and organizational change.

ENVIRONMENTAL STEWARDSHIP

Like most industries in the United States, the extractive industries have left a legacy of environmental problems. Early miners either did not understand the effects of their activities or believed that there was so much land available that it simply did not matter if some areas were damaged. Today, these adverse effects are seen as a problem that needs to be addressed.

Geography, geology, climate, and topography play critical roles in determining the type of waste produced and how mining can be conducted, thereby directly influencing the environmental consequences of the mining activity. Mining must, of course, be located where the mineral or other resource naturally occurs. The geology of the ore body or resource deposit dictates not only what target metals or resources are present, but also what nontarget or undesirable materials must be removed or disturbed during mining. Climate has direct effects on surface-water and groundwater hydrology and the management of mine drainage. In addition, temperature, winds, and other climatic factors influence how the mining can be conducted in a safe and environmentally responsible manner. Finally, topography affects not only hydrology and site access, but also placement of waste rock, processing facilities, and reclamation. Many of these constraints are unique to the extractive industries.

Early problems with coal mining, especially acid mine drainage, appeared first in the eastern United States. The proximity of population centers made these problems not only immediately apparent, but also likely to adversely affect large numbers of people. To address this legacy of poor environmental stewardship, Congress passed the Surface Mining Control and Reclamation Act (SMCRA) in 1977. SMCRA, a classic command-and-control approach to dealing with environmental problems, created a federal agency to oversee coal mining activities. Although SMCRA caused hardships among the coal mining industry, often forcing smaller operators out of business, it has virtually eliminated the abuses of the past. Further, through imposition of a tax on current coal production, SMCRA is generating the funds necessary to address long-standing problems. Typically, these funds are collected and used on a state-by-state basis. In a few states (e.g.,

Wyoming, where ongoing coal production from the Powder River basin is substantial), the funds recovered from current coal production have exceeded the amount needed to address historical coal-mining problems. This funding scheme provides some incentive for states to encourage new coal mining.

Hard-rock mining has also left a legacy of environmental problems, especially in the western United States. Acid drainage, habitat destruction, historical smelter emissions, and toxic waste piles are among the most prevalent negative impacts. Although the land area affected by hard-rock mines is only a fraction of 1 percent of total land area in the West, many of these former mining areas are now becoming more heavily populated. Other old mining areas are becoming popular vacation spots, and many are located near national and state parks and other recreation areas. As the visibility of these problem areas grows and population centers expand toward them, concerns about damage caused by mining and its effects on human health and the environment are heightened.

Every western mining state has passed legislation requiring currently operating mines to address these environmental concerns, and all new mines are required to reclaim the land they disturb. Some states have implemented programs to begin to deal with problems caused by past mining activities. However, there is as yet no comprehensive national system to address historical mine-waste issues. Indeed, there is disagreement over the scope and complexity of the problem. Some estimates indicate that there might be over 500,000 abandoned hard-rock mine sites in the United States. However, few of these sites pose any immediate threat to human health or the environment.

Modern mining companies for the most part recognize their responsibility to the environment and have adjusted their practices to avoid the problems of the past. Corporate codes of environmental practice and periodic environmental audits are now becoming standard practice throughout the industry. Standard environmental goals include

- zero-water discharges;
- minimization of air discharges;
- land reclamation that maintains, restores, or replaces site values (other than the value of the resource that is removed) whenever practical (in some cases, such restoration to original conditions is not practical and might even entail net environmental costs, such as the energy costs of backfilling open-pit mines);
- cleanup of abandoned historical sites or problems (e.g., idle oil wells, depleted hard rock mines, contaminated aquifers, acid mine drainage, or visual impacts);
- continual improvement in the classification and utilization of wastes; and
- continual improvement in the recycling abilities of smelters.

Improved environmental practices are being instituted not only in the extraction of the ore, but also in the processing of the ore to create raw materials for use in other industries. Consistent with an industrial ecology perspective, there has been a general shift from a linear-flow-of-production model to an ecosystem model. Leading firms begin environmental studies as soon as resources are discovered and apply design-for-environment principles to extraction, waste handling, and remediation plans from the earliest stages of project development.

One example is the treatment of waste rock that has the potential to generate acids by leaching. Strategies include segregating acid-generating rock from non-acid-generating rock; keeping air and water from the acid-generating rock; keeping air and water from the acid-generating material during storage (e.g., by covering or encapsulating); mixing the acid-generating material with other materials that have a large acid-buffering capacity; or placing the acid-generating material below the water table to prevent the oxidation of acid-generating sulfides. Some operations even take advantage of acid drainage to recover useful metals from the solution.

Expenses devoted to environmental protection are coming to be perceived as wise investments rather than financial sinkholes. In some cases, the investment value is quite tangible. In addition to avoiding future liabilities, such efforts can also provide income through the sale of pollution permits (M.J. Wilson, WZI Inc., personal communication, 1994). Spending for environmental protection might also lead to competitive advantage. For example, contractors working for the Europipe consortium building an oil pipeline under the Wadden Sea, a United Nations Biosphere Reserve on Germany's North Sea coast, found that abiding by environmental management standards improved the likelihood of their being hired on future projects (Grann, 1997).

Opportunities for recycling are limited in some of the extractive industries and common in others. For example, residues can be fed back to the concentrating, smelting, or refining operations, thereby recovering essentially all of the valuable metals. Indeed, some smelting operations produce slag with higher metals concentrations than the ore being mined. As a result, this slag is processed to recover those metals. More commonly, used batteries, metal drums, and used tires are returned to the manufacturers or suppliers for recycling. Used oil is often burned in approved and permitted waste-oil burners to recover energy as heat.

On a larger scale, many opportunities for reclaiming mines, including many abandoned sites, are being pursued. As population pressures have increased the demand for land, abandoned mines are being reclaimed for recreational, industrial, commercial, and even residential use. Many of the acid mine drainage problems of the past are being controlled through the use of conventional and innovative treatment systems. However, many of these end-of-pipe approaches have very high capital and operating costs.

Extraction of an ore body or resource deposit, like any industrial process, generates three basic outputs: products, by-products, and waste. These outputs are not always defined precisely. For instance, many of the by-products of extraction and processing have historically been considered waste. Coal-bed methane has typically been vented to the atmosphere, for example. Studies are under way to devise safe and efficient means of capturing and processing this methane to enhance energy supplies and reduce emissions. Used oil from mining equipment is being studied for use in blasting. Blasting operations normally employ a mixture of ammonium nitrate and fuel oil. If concerns over the fate of heavy metals in the used oil can be resolved, perhaps it could be substituted for the virgin fuel oil normally used.

Numerous opportunities exist for improving the energy efficiency of extraction operations. For example, conveyor transport of coal or ore and waste rock can be much more energy efficient than truck or rail haulage. Trade-offs between engine efficiency and emissions must be examined closely. Significant progress has been and continues to be made in this area.

Much work remains to be done to return mined land, particularly land at abandoned mines, to productive use. Approaches are needed that eliminate the generation of acid mine or acid rock drainage at abandoned mine sites. Use of artificial or enhanced wetlands to treat acid drainage is but one example of so-called passive technology that can address this problem. Reprocessing of historical waste using more efficient recovery methods can eliminate the source of acid rock drainage in some cases. Restoration of mined land—for wildlife habitat, parks, or other productive uses—must be encouraged.

As discussed above, outputs that have historically been considered waste could be utilized as by-products, if processing, transportation, and marketing barriers can be overcome. One key obstacle is the difficulty of communicating what by-product materials are available to those who might be interested in using them. Producers are often reluctant to provide detailed specifications of wastes or by-products for fear that competitors might gain an advantage from this knowledge. This reluctance can be minimized during resource extraction but is more difficult to overcome in the mineral-processing phase.

Another opportunity for extractive industries is to consider the development of a resource using a design-for-environment philosophy. For example, the use of more energy-efficient hauling and crushing operations provides direct benefits to the operator and also numerous indirect benefits such as reduced emissions of greenhouse gases. Similarly, energy recovery from waste-heat boilers connected with smelting operations offers opportunities to reduce energy requirements in mineral-processing facilities. Schemes to recycle or reuse water within the mining and processing operations reduce the demand on clean water supplies and the discharge of used water, which normally contains pollutants.

Further opportunities exist to maximize the recovery of by-products (and thereby minimize the generation of waste) through more efficient capture and segregation of by-product streams. More efficient capture of sulfur dioxide from smelters, for example, increases the production of sulfuric acid, which is the most common commercially traded chemical in the United States. Flue dust from smelters, which is often considered a hazardous waste, can contain very high concentrations of useful metals, and systems are being developed that allow the recovery of those valuable by-products.

A good deal of progress has been made in pollution prevention and waste minimization. Examples include the minimization of sulfur and ash from coal; improved means of disposal of toxic by-products of metals mining and processing; improvements in energy efficiency; more recycling of scrap; and better planning for the closure of mines (although the industry should beware of underestimating the costs of proper mine closure).

ENVIRONMENTAL REGULATIONS

Most current regulations are based on command-and-control procedures that assume linear streams of production and disposal. These procedures can inhibit innovation that would improve the environmental performance of the industry. For example, regulations can discourage interindustry assistance through bioremediation and other techniques due to company concerns about being named a potentially responsible party at a later date. They can also delay or even prevent remediation of acid drainage from old mines. For example, the National Park Service (NPS) has been trying to partner with a mining company to address historical mine waste and acid drainage on NPS land. No one has stepped forward, again because of concerns over legal liability. Initiatives now being developed under the guidance of the National Science and Technology Council might encourage such partnerships in the future.

Some environmental standards are excessively rigid or conservative. For example, the Environmental Protection Agency's (EPA's) cleanup standards regarding soil contamination are based on assumptions about how much soil children might ingest. However, regardless of how much dirt children eat, metals in mine tailings have low bioavailability; that is, they are mostly insoluble and simply pass through the digestive system. Thus, at least until recently, EPA's models overestimated bioavailability, and EPA consequently dictated maximum soil-metal concentrations lower than those that would be adequate to protect human health. These miscalculations have affected cleanup efforts at Aspen and Leadville, Colorado; Butte, Montana; Triumph, Idaho; and Salt Lake City, Utah.

The rigidity of current regulations does not allow cleanup practices to be tailored to particular sites. For example, in some cases, EPA's requirements dictate that the concentrations of metals in soil or water be reduced below the ambient concentration. Environmental protection dollars could be spent more cost effectively if background conditions were taken into consideration in decisions regarding acceptable contaminant levels at closed sites.

Environmental regulations associated with specific media (water, air, and land) are common in the extractive industries, as they are in other industries in the United States. To date, such regulations have been highly structured and narrowly defined, often specifying the use of particular technology to address medium-specific problems. The Clean Air Act, the Clean Water Act, the Hazardous and Solid Waste Act, the Resource Conservation and Recovery Act, and the Comprehensive Environmental Response, Compensation, and Liability Act are virtually all medium-specific and assume a linear flow of materials through industrial processes resulting in waste to the environment. This type of regulatory approach drives compartmentalized environmental thinking in industry rather than encouraging the systems approach characterized by industrial ecology.

In mineral processing, as in most conventional heavy industries, environmental stewardship has often consisted of end-of-pipe emissions controls. For pyrometallurgical (smelting) facilities, end-of-pipe treatment has typically relied on acid-control facilities and scrubbers to remove sulfur dioxide and metals from the off-gases. Acidic, metals-laden wastewater from the scrubbers is typically treated with lime to neutralize the acid and precipitate the metals. Similarly, wastewater from hydrometallurgical operations is often neutralized with lime.

These end-of-pipe solutions have the effect of transferring the environmental problem from one medium to another. For example, off-gases from smelting are scrubbed to remove metals and sulfur dioxide, but the problem is simply transferred from air to water, which is then treated to remove the metals and acid. The result is a sometimes hazardous solid waste that must then be managed.

EPA's enforcement of these laws has been similarly compartmentalized. The agency still believes that its policy of "enforcement first," which is typically carried out one law at a time, will yield the best results. Although this command-and-control, medium-specific approach has generated some successes, it has also stifled innovation. In addition, because some of these programs are delegated to states, and often the regulations are enforced by different agencies with different priorities, the costs of complying with regulations add up very quickly. EPA's approach has often led industries to shift the environmental problem from one medium to another, without eliminating the real source of the problem.

EPA has recently announced that it will assemble multimedia inspection teams to examine environmental impacts in water, air, and land at a particular site. This is a good first step toward considering environmental effects holistically, but the approach should not be limited to enforcement situations. Flexible methods for determining the relative risks and benefits of the traditional command-and-control approach versus a more voluntary, incentive-driven approach must be explored. Use of public education and pressure, for example, can be a very effective method for encouraging industries to reduce the amount of toxic materials they release into the environment. Perhaps one of the best success stories is the toxic release inventory, in which EPA publishes the amount of toxic materials released to the environment from companies in various industries. Companies have a strong public relations incentive to not be ranked high on this list, and many firms have publicly committed themselves to specific reduction goals. To encourage innovative approaches to minimizing waste, there must also be incentives for industry to take risks without fear of heavy-handed enforcement if their innovative solutions do not meet or exceed established goals.

Future environmental rules and regulations should provide a better balance of carrots and sticks. The industry would have more incentive to take voluntary action if the regulators gave credit for such actions. For example, Kennecott Utah Copper's modernized smelter will have emissions levels far below those required by regulations. When it starts up, the permissible emissions rate for sulfur dioxide will be 3,200 lb/h, but the actual emissions rate will be less than 200 lb/h. However, because final rules have not yet been promulgated for nonferrous smelters, federal agencies have been reluctant to give Kennecott acid rain credits for its excess reduction of sulfur dioxide emissions. (Final rules have been promulgated for electric power utilities, permitting companies to take credit immediately for sulfur dioxide reductions.)

Other examples of voluntary actions include Kennecott's cleanup of lead tailings off Kennecott property along Bingham Creek in Utah (neither Kennecott nor any of its predecessors mined or milled lead ores); reclamation of hundreds of acres of waste-rock dumps, even though not required by any regulation; and demolition and reclamation of obsolete facilities. None of these actions was called for by regulators, nor was any credit given for exceeding requirements.

Regulations that are sensitive to the characteristics of particular sites would allow a more efficient allocation of funds to environmental protection and remediation efforts. Rather than reducing contaminant concentrations below ambient levels at one site, for example, funds could be invested more productively elsewhere. In addition to being sensitive to the background chemistry of a site, regulations should be sensitive to other site features as well, such as grade, impurities, geometry of the ore body, and climate. For example, a mine that generates waste rock that is not acid producing would not need extensive water protection efforts, because acid drainage would not be a factor (Box 1). Similarly, a mine generating acid-producing waste rock in a very dry climate would not need the same extent of water-protection efforts as a similar mine in a wet climate.

Greater efforts need to be made to allocate efficiently funds for environmental protection by ensuring that the regulations are scientifically defensible, not excessively conservative. In many cases, interdisciplinary discussions might be necessary to achieve consensus regarding appropriate standards.

Regulations need to distinguish between past practices, existing operations, and new developments. It might make sense to apply separate sets of regulations for new and existing mines and another set for past practices. The rationale for this approach is based on cost-benefit analysis: The cost of avoiding a problem might be only a small fraction of the cost of correcting the problem. If the notions that benefits should somehow be related to costs and that funding is limited (or that funds could be spent to greater effect if used elsewhere) are accepted, then it might make sense, from a societal view, to accept some degradation in limited areas.

Although past practices have left a legacy of environmental problems, most

> **BOX 1 Acid Drainage**
>
> Acid drainage forms when sulfur (present in ores in the form of metal sulfides such as pyrite and FeS_2) reacts with oxygen and water to form sulfuric acid. This acidic solution can dissolve other minerals, thus leaching metals. A number of environmental problems are associated with this phenomenon, which also occurs naturally whenever sulfur-containing rocks are exposed to air and water. These include toxicity to plants and (mostly aquatic) animals, resulting not only from depressed pH (acidity), but also from the dissolution of toxic metals. Acid drainage can also cause physical habitat degradation, because when natural processes neutralize excess acidity, hydroxide precipitates are formed that can smother benthic organisms. Acid drainage can also cause visual pollution by staining streams and lakes. All of these problems stem from the initial acid-forming reactions.

present operations are environmentally responsible and in compliance with existing rules and regulations. The most pressing environmental concerns of the industry are that

- environmental regulations be based on sound science;
- environmental regulations take into account local circumstances;
- development of known reserves remains economically feasible; and
- strategies developed for remediating historical problems do not threaten the short- or long-term viability of current and future operations.

Addressing these concerns effectively will require the mining industry to convince the public and government that current operations are environmentally responsible. Doing this will, in turn, depend on effective communication and public education.

LIFE-CYCLE PRACTICES

Although recycling is common in some segments of the extractive industries, little in the way of formal life-cycle assessment (Box 2) is being done, mainly because of where materials extraction occurs in the life cycle of finished products. Making inventories of the energy and resources necessary to extract and process raw materials is relatively easy. However, should a life-cycle assessment (LCA) that evaluates the use and disposal of the raw materials be performed, given that the raw materials might have many uses?

Even when the industry is viewed as a system unto itself, LCA is fraught with difficulties. For example, coal seams in the eastern United States are typically

> **BOX 2 Life-Cycle Assessment**
>
> An environmental life-cycle assessment typically involves
>
> - taking stock (inventory) of all materials and energy inputs and outputs (including wastes) and associated environmental loadings of a product;
> - analyzing the impacts of the environmental loadings identified in the inventory; and
> - evaluating opportunities to reduce the environmental burden associated with the life cycle of that product.

thin, and large areas of land must be disturbed to gain access to the coal. This coal is often high in sulfur content. Western coal, in contrast, is more often found in thick, continuous seams and has a lower sulfur content. However, reclamation of western lands can be more difficult because precipitation rates are lower, and the coal must often be shipped long distances for use in power plants. Is it the responsibility of the coal producer to consider these reclamation and transportation costs? How are the differences in sulfur content accounted for? Some power plants are equipped with scrubbers to remove sulfur from the off-gases, but these scrubbers typically produce large volumes of sludges that must be disposed of.

When the hard-rock mining industry is examined, the complexities are even greater. Metals such as copper, zinc, and nickel are produced worldwide, and prices for these commodities are set on the world market. Inherent properties of the ore body impose constraints that cannot be easily overcome, particularly with respect to nontarget metals such as arsenic or mercury.

In the mineral-processing area, more traditional LCA approaches can be used. Mining and petroleum companies have an interest in the fate of their products after they are sold, but they are most interested in extraction and primary processing. Thus, the life-cycle concept is most usefully applied to the life of the mine or oil field, not the life of the product.

To be useful for the extractive industries, LCA must be able to carefully describe the system. This requires appropriate metrics. For coal extraction, for example, one appropriate metric might be the ratio of the amount of energy needed to extract a unit of coal to the amount of energy that unit can provide. Although such a metric might be appropriate when the system to be considered is the coal mine itself, it is probably not appropriate when the amount of energy needed to ship this coal and the production efficiency of the generating station are considered.

For a system defined as the processes and facilities needed to deliver energy to an end user, a better metric might be the energy required to deliver a unit of energy to the end user. This measure would take into account not only the energy used in mining the coal, but also, for example, the energy used to transport the coal and convert the coal into electricity and the energy lost in conveying the generated electricity to the end user. The issue here is: Who defines the appropriate system? Certainly, a coal producer cannot be expected to cope with the complexities of rail transport, power generation, and power transmission, let alone the complex social and economic issues that surround siting rail lines versus power transmission corridors or disposal of ash and scrubber residue.

Metals extraction poses similar systems definition and measurement challenges. If the system is defined as the mine itself, inherent properties of the ore body will dictate most of the characteristics associated with waste generation. Stripping ratios (i.e., the amount of overburden, or waste rock, that must be moved to expose a given quantity of ore), ore grades, accessory mineral content, and other factors determine the quantity and composition of the waste rock. Although in some cases the stripping ratios can be controlled (e.g., through consideration of underground versus open-pit mining technology), in most cases the constraints imposed by the geometry of the ore body will dominate. Nevertheless, metrics can be defined that provide meaningful measures of mine performance. As for the case of coal, above, the amount of energy needed per unit of metal extracted might be a good metric.

In ore processing, more conventional metrics might be appropriate. For example, 3M currently uses a metric that is the ratio of the mass of waste to the combined mass of products, by-products, and waste. This single figure provides a simple measure not only of the degree to which waste can be reduced, but also the degree to which waste can be converted to by-products for recovery and sale. This type of metric also could be applied to the processing of ores.

Regardless of the metric chosen, it must be easy to calculate based on data that can be readily obtained. Ideally, the metric should be a single index or number that can be used to establish goals and measure progress toward those goals. It must at least be suitable to the variations of the particular industry being measured and, ideally, comparable across industries.

As discussed above, the nature of materials extraction constrains the application of life-cycle concepts. In addition, the remote locations of most resource deposits, combined with the low value of mining waste, limit the potential for selling waste products to other industries. Because the managers of extractive industries have limited control over the fate of their products, it might be more productive for them to think in terms of the life cycles of sites (e.g., mines or oil fields), rather than life cycles of products. Rather than dwell on the difficulties of applying all industrial-ecology approaches to extractive industries, the objective should be to apply design-for-environment principles to the mining and processing phases and to maintain or restore the other values of the site after the resource has been extracted.

CULTURAL AND ORGANIZATIONAL CHANGE

Perhaps more so than other industries, mining is a global enterprise. Particularly for hard-rock mining, the marketplace for commodities is worldwide. Indeed, the prices of most metals are set on the world market. In addition, the prices of metals are subject to radical fluctuations. The need to maintain global competitiveness in this atmosphere requires that hard-rock mining companies adjust rapidly to these fluctuations and inhibits their ability to plan for the long term. As a result, some mining companies have suffered the consequences of short-term thinking. Even today, a few mining companies do not consider such basic needs as planning for ore depletion and ultimate mine closure.

Hard-rock mining, and especially open-pit mining, necessarily defers the bulk of mine-closure expenses to the end of the mine's life. This involves establishing large bonds or other financial assurance mechanisms to ensure that adequate funds will be available (in light of volatile metal prices) to close the mine once the economically recoverable reserves are exhausted.

Very large mining companies are able to absorb or weather the consequences of rapidly fluctuating commodities prices and are therefore able to plan for the long term. A diversity of properties and metals also enables large companies to deal with end-of-mine-life closure costs for open-pit operations. Consolidation of smaller mining companies into a few larger organizations appears inevitable for these reasons.

When viewed as a closed system, the hard-rock mining industry has limited recycling opportunities (except, perhaps, in its ore-processing operations). However, when viewed in a larger context, recycling opportunities for metals are much greater. For example, as shown in Table 1, a much greater degree of recycling appears at least theoretically possible for a wide variety of metals (Allen and Behmanesh, 1994).

Yet, as discussed earlier in this paper, recycling in some senses competes directly with extraction of virgin materials. The market costs of recycling are often much greater than the cost of producing virgin materials, and there can be significant direct and indirect adverse effects associated with recycling (e.g., excessive energy used to collect the dispersed materials, or pollution generated from burning insulation off copper wires). Until there is an economic incentive for society to prefer recycled materials over virgin materials, recycling will not be viable.

There are several ways to provide incentives to recycle rather than produce virgin materials, but these incentives will have to be society driven rather than free-market driven. This is because society will have to absorb the increased costs associated with recycling. If the "external" costs of producing virgin materials, such as loss of usable land or other forms of environmental degradation, are somehow taken into account, the extractive industry will pass these costs on to the users of the commodities. In the global metals market, these external costs must be applied on a global basis. However, there is no commonly accepted

TABLE 1 Metal Recycling

Metal	Percent Theoretically Recoverable	Percent Recycled in 1986
Antimony	74–87	32
Arsenic	98–99	3
Barium	95–98	4
Beryllium	54–84	31
Cadmium	82–97	7
Chromium	68–89	8
Copper	85–92	10
Lead	84–95	56
Mercury	99	41
Nickel	100	0.1
Selenium	93–95	16
Silver	99–100	1
Thallium	97–99	1
Vanadium	74–98	1
Zinc	96–98	13

SOURCE: Allen and Behmanesh, 1994.

system for estimating what these costs might be, let alone how such a system might be implemented.

Similarly, the external costs of disposing of materials (e.g., contamination from land disposal and subsequent groundwater or surface-water contamination) might also be considered. Again, however, there is no commonly accepted method for calculating these costs or applying them globally. To add to the complexity, a balanced system of assessment must also take into account the external costs of recycling. To accomplish such a goal, collaboration between industry and government—not just nationally, but globally—will be essential.

There is a trend toward full accounting of environmental costs, but this is not a simple matter in the extractive industries. There is no practical means of incorporating the consequences of exhausting finite global resources in the calculation of environmental costs. Yet, only by explicitly identifying potential environmental costs before projects begin can anticipated consequences be managed and designed for. The trend toward establishing comprehensive background environmental data will assist with post hoc efforts to distinguish impacts due to the extractive activities from those caused by natural environmental changes.

Many of the difficulties faced by the extractive industries stem from inaccurate public perceptions of the industries' environmental performance and from misconceptions about the relative risks of various activities. These inaccurate perceptions are based mainly on historical practices and thus do not reflect recent improvements in the industries' environmental performance, and they represent a failure on the part of these industries to communicate effectively with various stakeholders. A variety of case studies demonstrate the benefit of direct, open communication. For example, as a result of extensive communication efforts, Statoil was given permission to build its Europipe for delivering oil from the North Sea to Germany under the Wadden Sea, a rich natural area that is both a German national park and a United Nations biosphere reserve (Grann, 1997).

Long-term communication efforts should focus on educating representatives of government agencies and, especially, the public. Regulators, teachers, and members of the news media are particularly suitable leverage points for education efforts, the objective of which should be to improve literacy about environmental issues in general and about relative risks in the extractive industries in particular.

Industry representatives will not be able to accomplish this task alone. Society as a whole will benefit most if, in addition to industry efforts, university practices and curricula are transformed. A better appreciation of the concepts of industrial ecology and the efforts of industries will depend on exposing students not to single disciplines, but to interdisciplinary studies sculpted by the nature of practical problems. Two stellar examples of innovative curricula are the earth systems course at Stanford University and the earth resources program at the University of California at Berkeley.

The traditional adversarial relationship between industries and government regulators can inhibit improvement in environmental performance. Accumulating examples demonstrate the potential of government-industry cooperation, however. The development of the clean-coal technology program and the joint Amoco/EPA study of Amoco's Yorktown, Virginia, refinery are two such efforts (Amoco and U.S. Environmental Protection Agency, 1993). However, a shift toward more cooperation will require cultural change not only on the part of industry, but also on the part of Congress and the regulatory agencies.

REFERENCES

Allen, D.T., and N. Behmanesh. 1994. Wastes as raw materials. Pp. 69–89 in The Greening of Industrial Ecosystems, B.R. Allenby and D.J. Richards, eds. Washington, D.C.: National Academy Press.
Amoco and U.S. Environmental Protection Agency (EPA). 1993. Amoco–U.S. EPA Pollution Prevention Project—Yorktown Virginia. Executive Summary. Washington, D.C.: EPA.
Grann, H. 1997. Europipe development project: Managing a pipeline project in a complex and sensitive environment. Pp. 154–164 in The Industrial Green Game: Implications for Environmental Design and Management, D.J. Richards, ed. Washington, D.C.: National Academy Press.
Wilson, M. J. 1994. WZI, Inc. Personal communication.
World Commission on Environment and Development. 1987. Our Common Future. New York: Oxford University Press.

Primary Materials Processing

CHARLES G. CARSON III, PATRICK R. ATKINS, ELIZABETH H. MIKOLS, KENNETH J. MARTCHEK, AND ANN B. FULLERTON

SUMMARY

In industrial ecology, industry sectors and various production processes are viewed as interconnected systems. This view extends environmental strategies beyond individual companies to address the integrated nature of economic activities. An analytical assessment of materials and energy flows forms the basis for improving the overall economic performance of a firm and of the integrated systems of industrial activities in which the firm plays a role.

By transforming and recasting materials and recovering embedded energy, the primary materials processing industry (PMPI) plays a unique role in industrial ecology and in the economy. This industry sector acquires raw materials from mining operations. Some PMPI companies add value to mined ores; others add value to sand and stone. Their products (e.g., copper, aluminum, steel, and cement) are used elsewhere in the economy to make such things as wires, cans, and construction materials. Many PMPI companies recycle and reprocess end products as part of their operations. Their place in an industrial ecosystem does not end there, however. They also use by-products and residues from other industries in their base processes. For example, cement kilns recover energy by combusting waste and use waste (e.g., mill scale, foundry sand, slag, or fly ash) from other industries in cement production. The aluminum industry utilizes low-value materials such as petroleum coke from refiners and coal-tar pitch from coke ovens to form electrodes for aluminum smelting.

The steel industry recycles mill scrap, fabricator waste, automobiles, structural steel, appliances, pipes, industrial machines, and tin-coated steel cans. In fact, steel is recovered for recycling at a higher rate (by weight) than any other commodity. Sixty-three million tons of steel were collected for recycling in 1993 out of the 100 million tons produced or imported (American Iron and Steel Institute, 1993).

Likewise, recycling is an important part of the aluminum industry. In some applications, such as packaging and automobiles, aluminum is recycled at a higher rate than steel. Over the past 20 years, aluminum-can recycling has taken off. In 1993, 2.9 million tons of aluminum from cans were recovered in the United States. It takes 95 percent less energy to recycle an aluminum can than to produce a new one, making recycling highly cost-effective from a business perspective. Fifty-three percent of today's beverage cans and over 60 percent of the aluminum used in automotive applications are made from recycled metal. In 1993, 63 percent of cans and 85 percent of aluminum automobile scrap were recovered and productively reused (Aluminum Association, 1993).

Energy is used to transform mined material into usable product just as energy is used to produce and deliver electricity, gas, and petroleum. The energy intensity of primary materials processing is exceeded only by the energy used to make energy itself. Energy efficiency improvements, therefore, are important to PMPI. The sector also can and does reduce energy use by burning waste from other industrial sectors in its high-temperature furnaces.

Long considered smokestack industries, PMPI companies handle much of their pollutants through recycling, reuse, and pollution control technologies, which have been put in place largely in response to environmental regulations. More recently, efforts have been made to eliminate pollutants at the source. For instance, the aluminum industry is currently reducing its production of trace polyfluorinated carbon gases in smelting by improving process monitoring and operating practices.

The growth in environmental regulations over the last 25 years has been paralleled by increasing societal demand for cleaner industries and improved environmental performance. In the past decade, global competition for PMPI products has intensified. Like many other industries, PMPI has contained or cut costs to remain profitable. The new set of linked economic and environmental challenges is being met by managers increasingly conscious of the cost, quality, and environmental impacts of their actions.

From an environmental perspective, PMPI companies face the following challenges:

- integrating environmental considerations in business decisions so that they can move beyond day-to-day regulatory compliance to true environmental stewardship;
- dealing with regulations and standards that are not keeping pace with changing practices;
- using environmental life-cycle practices internally to assess and reduce energy and materials use in their operations, even though methodologies have not yet been standardized;

- linking total quality management concepts with environmental decision making to reduce waste, eliminate unnecessary processes, and improve product acceptability and usefulness;
- guarding against the premature adoption of life-cycle assessment for regulatory purposes;
- leveraging external research and development (R&D) resources to offset reductions in internal R&D spending; and
- developing collaborative links (partnerships) with government regulators, environmental interest groups, and universities to address current and future environmental concerns.

ENVIRONMENTAL STEWARDSHIP

Today, companies often frame their environmental efforts in terms of environmental stewardship. According to the dictionary definition, stewardship is the "careful and responsible" management of operations and property within one's care. Most companies' reports suggest an increased regard for the environment, prompted by the growth in environmental regulations and voluntary and incentive schemes for pollution prevention. The environmental efforts of companies are also driven by economic considerations (such as meeting customer demands) and changing engineering and management practices (such as integrated product development and total quality management).

In the 1970s and 1980s, controlling emissions from processes and facilities was considered sufficient environmental care. In the 1990s, the emphasis is on preventing pollution and on isolating and reducing the environmental impacts of products and processes throughout their life cycles. The life-cycle perspective extends beyond the factory gate to include waste products that might be sent to a landfill or incinerator, materials and products that are sold, and materials and components that are purchased as inputs to products and processes. For management, this perspective means taking into account the practices of one's suppliers and managing the supplier chain. Management also needs to consider the environmental impacts of finished products once they are purchased by a customer.

Efforts to control pollution and reduce waste through prevention have proved fruitful. Environmental gains achieved in the 1970s at U.S. Steel (Box 1) illustrate the effectiveness of control technologies and prevention techniques in reducing environmental impacts. At the same time, other practices that were considered environmentally proper in the 1970s and 1980s are being reevaluated. For example, hazardous waste from the chemical industry and other industries is burned in cement kilns, an approach that reduces the need for other treatment technologies and provides energy to fuel the kilns. Although this practice is considered acceptable and proper in Japan (Box 2), in the United States, differences in regulations affecting these kilns and hazardous-waste incinerators are raising

> **BOX 1 Environmental Stewardship in the 1970s: Pollution Control Technology for the Coke Production Process**
>
> The coke production process (i.e., the carbonization of coal) is inherently dirty, and many pollutants are emitted. In the 1970s, U.S. Steel began installing new pollution-control technology, improving operating and maintenance practices, and developing and implementing an employee training program. The results from its Clarion, Pennsylvania, facility are striking: Benzene air emissions have dropped from 590,000 kg/yr to less than 27,000 kg/yr over the 1990–1993 period; sulfur dioxide emissions declined from 0.05 ppm to about 0.015 ppm; and particulate emissions average less than 25 $\mu g/m^3$ on a monthly basis. In addition, during the same period, the plant successfully reduced the total volume of solid waste requiring disposal from about 9,000 to 6,000 tons.
>
> SOURCE: Carson.

concerns (Box 3). Currently in the United States, approximately half of hazardous-waste solvents are burned in cement kilns, with the remainder being consumed in on-site boilers and incinerators or commercial waste incinerators.

In the late 1980s and early 1990s, based on anticipated increases in waste volumes, the U.S. waste industry invested significantly in permitted hazardous-waste incinerators. These actions were spurred by the Environmental Protection Agency's (EPA's) policies intended to increase the country's capacity to manage hazardous waste. At the same time, EPA began encouraging pollution prevention, and industries began implementing waste-reduction policies. The combination of industrial waste reduction and construction of hazardous-waste incinerators led to an excess of hazardous-waste incinerator capacity.

The incineration industry contends that a wider variety of waste should be steered to incinerators built specifically to handle hazardous waste and that the "less-regulated kilns" should come under greater regulatory scrutiny. The implication of this argument is that boilers, furnaces, and cement kilns are generally operated under less-rigorous environmental requirements than are commercial hazardous waste incinerators. Under EPA's 1991 rules for boiler and industrial furnaces, however, cement kilns handling hazardous waste must test each batch of waste to be combusted, continuously monitor toxic-metal feed rates, demonstrate destruction of organics, and continuously monitor emissions (all the while ensuring that the facility makes a quality product). In some cases, these requirements exceed the standards that hazardous-waste incinerators must meet.

**BOX 2 Japan's Use of the Cement Industry
for Recycling of Industrial Waste**

For the past 20 years, the Japanese cement industry has been successfully recycling many industrial waste products from a variety of industries. This activity has served two goals: overall industrial waste reduction and increased energy efficiency for the cement industry.

The table below outlines the various industries and their respective wastes that are used as resources for the Japanese cement industry. Specific statistics indicate that currently 60 percent, or 15.6 million tons/yr, of Japan's blast-furnace slag is recycled in cement kilns. Three-quarters of the fly ash produced in coal power plants, about 1.6 million tons/yr, is recycled as cement, and the rest is sent to landfills. Half the used tires generated annually in Japan, about 400,000 tons worth, are recycled, and 37 percent are used as a source of energy. Of the latter, 40 percent are burned in cement kilns. Bota, huge heaps of coal waste once common in coal-mining areas, are gradually disappearing as they are consumed by the cement industry.

Industry Source and Type of Waste Used in Japanese Cement Industry

Source	Waste Used
Power generation	
Coal-fired	Fly ash from coal-fired plants, stack gas desulfurized gypsum
Crude-oil-fired	Fly ash from oil-fired plants, stack gas desulfurized gypsum
Coal mining	"Bota" (coal waste)
Steel refining	Blast furnace slag, pig iron furnace slag, electric furnace slag, converting furnace slag
Nonferrous refining	Copper slag, iron concentrate, stack gas desulfurized gypsum
Metals manufacturing	Casting sand waste, waste wire covering
Oil refining	Oil cokes, catalyst residue, used kaolin
Chemicals	Automobile tires, waste paint, waste oil, stack gas desulfurized gypsum
Papermaking	Paper sludge, incineration ash of pulp
Fuel oil	Used kaolin, waste oil
Sugar manufacturing	Waste sugar dregs
Beer brewing	Used diatomaceous earth
Construction	Waste earth from construction

SOURCE: Akimoto, 1994

> **BOX 3 U.S. Use of Cement Kilns for Hazardous Waste Treatment**
>
> Since the mid-1970s, cement kilns in the United States have used a variety of hazardous wastes as supplemental fuel. Until recently, this waste tended to be solvents, other cleanup materials, and by-products created during organic chemical processes. Such wastes are generally in the form of a pumpable liquid, often with small particles such as paint skins, fibers from rollers, and other small-particle-size sludges and contaminants. The liquids have an energy value from 8,000 to 13,000 Btu/lb, similar to the energy value range of coal, the primary fuel for a Portland cement kiln. Recently, sludges and solids with a wider range of physical and chemical composition have been used as an alternative fuel. Instead of being pumped through a nozzle and burned with the coal, these sludges and solids are often introduced at the midpoint of the kiln. They can also be ground and emulsified with the liquid waste.
>
> In the early 1980s, EPA sponsored a number of air-emissions experiments at cement kilns that burned fuels containing hazardous waste. The results showed that the cement kiln can adequately combust even the most stable organic compounds with destruction and removal efficiencies exceeding 99.99 percent. This level of combustion efficiency is possible because of the physical nature of the kiln itself. Because the manufacture of Portland cement clinker requires a minimum reaction temperature of approximately 2550°F, the flame temperature is usually between 3000°F and 3500°F. The size of the kiln, which is several hundred feet long, provides a sustained residence time for exhaust gases. Its rotary action, required to move the raw materials to the burning zone, where the final reactions take place, and finally into the clinker cooler, provides a turbulent atmosphere within which complete mixing occurs. These test results, coupled with the capacity of combustion facilities for the disposal of haz-

This debate does not take into account the different functions of the two systems, as summarized in Table 1. Kilns, boilers, furnaces, and waste incinerators perform different services (incinerators incinerate hazardous waste and are reliant on a continuous feed of waste, whereas cement kilns recover energy from waste to make a useful product, cement) and handle different types of waste. Generally, boilers and industrial furnaces require waste with a significant energy value. In contrast, incinerators can take waste with no energy value. Incinerators are also more able to receive and handle wastes of mixed physical consistency, whereas boilers and furnaces preferentially use pumpable waste. Industrial furnaces might sometimes be limited in the types of wastes they use because of chemical criteria imposed to maintain product quality.

ardous-waste liquids and with the positive environmental benefit of recovering energy from waste, led EPA to conclude that cement kilns could safely destroy organic waste contaminated with small amounts of metals. EPA decided that kilns and other industrial boilers and furnaces could continue to consume hazardous waste under some circumstances until such time as EPA prepared and issued the regulations needed to control this activity. The facilities would eventually be required to obtain full permits for hazardous-waste treatment, storage, and disposal operations under the Resource Conservation and Recovery Act.

In early 1991, EPA issued regulations for boilers and industrial furnaces burning hazardous waste. The regulations require that cement kilns, a class of industrial furnace, satisfy health-based standards when burning hazardous waste. The kilns must control emissions of particulate matter, heavy metals, dioxin and furan, hydrochloric acid, and organics. The regulations also require extensive waste-analysis plans certifying the quality of burned waste, spill-prevention plans, health and safety training plans, specific operating conditions, as well as plans for other operations aimed at protecting the environment.

From the cement industry's perspective, these regulations exceed those for hazardous-waste incinerators and do not take into account the environmental benefits of cement kilns over incinerators for treating hazardous waste. The cement industry is also concerned that similar levels of rigorous waste analysis and emissions testing might be required for recycling of other nonhazardous waste in cement kilns, thus impeding recycling at these facilities.

SOURCE: Mikols.

As the cement industry has made its case for the advantages of treating waste in cement kilns, it has learned firsthand the importance of effective communication. Most PMPI companies recognize the need to improve their environmental communications, if they are to be successful environmental stewards. Communication is key to implementing new environmental practices within a company, helps educate regulators and communities in which these firms do business, and engages nongovernmental environmental organizations in discussing the scope and complexity of environmental issues and potential solutions. Alcoa's efforts to take into account community views of the environmental life-cycle impacts of operating and closing a bauxite mine in the Jarrah Forest in Australia (Box 4) illustrate effective environmental communications in action.

TABLE 1 Comparison of Hazardous Waste Incinerators and Cement Kilns

	Typical Cement Kiln	Typical Hazardous Waste Incinerator
Maximum gas temperature	>3500°F	≤2700°F
Maximum solid temperature	2600°F–2700°F	≤2500°F
Gas retention times at ≥2000°F	3–10 s	0–3 s
Solid retention times at ≥2000°F	20–30 min	2–20 min
Turbulence (Reynolds number)	≥100,000	≥10,000
Size/Speed	180–600 ft long 10–25 ft diameter 1.0–2.0 rpm[a]	15–60 ft long 10–20 ft diameter 0.5–2.0 rpm
Loading	Typically 5–10% of input is waste Raw material 70–250 tons/h Coal/coke 5–10 tons/h Waste fuel 5–10 tons/h	100% of input is waste Waste 5 tons/h
Volumes of air handled, gas flow rate	Average: 120,000–130,000 dscfm[b]	Average: 25,000–30,000 dscfm
Raw material processing	70–250 tons/h	None
Viable product produced	Cement clinker	None
Conservation of fossil fuels	Approximately 20% of cement kilns use hazardous waste as fuel. The replacement rate can be between 30 and 100% of traditional fuels, depending on individual plan circumstances. This amounts to a savings of about 1 million tons per year.	Incinerators rely on the generation of hazardous waste and thus provide no resource conservation.

[a]rpm = revolutions per minute.
[b]dscfm = dry cubic feet per minute adjusted to standard conditions.
SOURCE: Mikols.

Intraindustry collaborative efforts have also become an increasingly important aspect of environmental stewardship in the 1990s. Companies competing in the same industry share common environmental concerns. This commonality has led to formal and informal exchanges of nonproprietary technology and information, such as the aluminum industry's current efforts to reduce polyfluorinated

BOX 4 Community Involvement in Bauxite Mining in Australia's Jarrah Forest

Bauxite is one of the basic raw materials for primary aluminum production. Reserves of it exist in the Jarrah eucalyptus forests that are unique to Western Australia. Over the past several years, Alcoa's Australian subsidiary has developed a comprehensive approach to mining bauxite in this environmentally sensitive area south of Perth.

Before extraction at a new site, 5- and 10-year mining plans, compiled in conjunction with government agencies, are prepared. These plans take into account all environmental considerations in the life cycle of the mine, including its operation and rehabilitation. The objective is to restore the land to a condition that serves the forest products industry, provides recreational opportunities, serves as a water catchment for the city of Perth, and provides habitat for native wildlife.

During the mining process, timber is harvested and used by lumber producers and wood chippers. Other vegetation is burned, and the ash is used on site to preserve nutrients. Topsoil from the active mine area is moved immediately to areas under rehabilitation. Research indicates that storing topsoil adversely affects the seeds, spores, and nutrients contained in it. By retaining the ash on site and reusing the topsoil immediately in other mined areas, however, biodiversity levels approaching those of the native forest can be achieved quickly once the mine is closed.

This approach differs in two important ways from past practices. First, it takes a life-cycle view. Second, it necessitates a collaborative effort with the surrounding communities, providing them with information and education on the scope of the environmental issues and the pros and cons of different solutions.

Since 1983, over 100,000 people have visited the Alcoa mines on a tour designed to educate the public on reclamation activities. The Jarrah Deeback Research Centre is also open to the public. Public input and dialogue are sought on issues such as land use, recreation, blasting, and transportation.

Alcoa also sponsors a community land-care project, a $6.5 million, 5-year effort in support of Australia's National Decade of Landcare. This community assistance program is designed to put the company's resources and its operational experience at the disposal of community land-care initiatives. In Western Australia, where $5 million has been budgeted for the project, the land-care program includes

- agricultural subcatchment demonstration sites in the wheat belt;
- a wetlands rehabilitation project;
- a land-care field study;
- a comprehensive range of education, information, and community awareness programs and recreational facilities; and
- expanded support for the Greening Western Australia movement.

SOURCE: Atkins.

> **BOX 5 Intraindustry Collaboration to Reduce Pollutants from Aluminum Production**
>
> Aluminum production by the Hall-Heroult electrolytic process involves the use of carbon anodes submerged in a relatively high-resistance eutectic containing fluoride salts. If resistance becomes too high, the cell voltage rises, and an anode effect occurs. This results in the production of polyfluorinated compounds such as carbon tetrafluoride (CF_4) and hexafluorethane (C_2F_6). Both gases are extremely stable in the atmosphere, with lifetimes estimated to be greater than 1,000 years. They also have a high absorption potential for infrared radiation and are considered strong greenhouse gases. These characteristics have made them pollutants of concern.
>
> The worldwide aluminum industry is addressing this issue by developing improved measurement techniques, establishing emissions inventories, studying the relationship between anode effect and PFC emissions, and designing and implementing electrolytic cell operating systems that significantly reduce the frequency and duration of anode effects.
>
> In March 1994, an international workshop on the measurement and management of PFCs, particularly their potential impact on global climate, prompted U.S. industries to establish a voluntary agreement with EPA to reduce significantly PFC emissions.
>
> SOURCES: Atkins and Martchek.

carbon (PFC) emissions (Box 5). Such collaborative efforts appear to be a viable mechanism for moving beyond compliance to pollution prevention. For example, the U.S. aluminum industry is nearing an agreement with EPA to reduce voluntarily by 50 percent emissions of PFCs at their source.

As the PMPI sector continues to improve its environmental stewardship, it faces several challenges:

- working with regulators to ensure that environmental requirements are based on good science and are economically and technically feasible;
- developing mechanisms for working collaboratively with all stakeholders to reach mutually acceptable solutions; and
- implementing pollution-prevention practices in the midst of regulations and standards that are still based on a command-and-control model.

REGULATIONS AND STANDARDS

National and international environmental regulations and standards focus primarily on controlling pollution and handling waste from industrial and societal activities. These statutes and standards have helped advance air and water pollution technologies, landfill design, land remediation technologies, and the development of monitoring devices. Companies have to comply with existing and new regulations or face stiff penalties. Few will disagree that pollution controls are necessary. However, several existing regulations and standards promote control of pollution rather than its prevention. They also impede the creation of industrial ecosystems that minimize waste in clusters of industrial systems or sectors. Part of the problem is the way in which these rules define waste. To change the regulatory mechanisms so that more materials find useful purposes instead of landfill space, the definition of waste might need to be reevaluated.

Materials are defined currently by their fate. Product or process inputs are "useful" materials. Materials for which no useful application has been found are considered "waste." The fate of materials emerging from industrial processes is dictated increasingly by regulatory definition. Pollution control regulations based on the paradigm of linear flows of materials in turn promote linear flows of materials through the economy. Materials emerging from industrial systems are therefore defined as either product or residual material. For example, language in the Resource Conservation and Recovery Act (RCRA), which focuses on the disposal and treatment of waste, sometimes results in recyclable and reusable material being considered waste (Box 6).

Regulations based on a linear materials flow model lead to two assumptions about waste: It has no economic value, and it is of inferior quality. By treating recyclable and reusable material as waste, RCRA inhibits efforts to minimize

---

BOX 6   Potlining: Waste of a Reuseable Residual Material

Potlining is a residual material produced by the aluminum industry. It contains fluoride and cyanide. When used in cement manufacture, the cyanide is destroyed, and the fluoride is used beneficially in the cement. Regulations consider potlining to be a solid waste rather than a raw material. This regulatory definition of a potential raw material as waste led Santee Cement Company to stop using potlining in its process to avoid the costs of obtaining permits. Thus ended the potential reuse of 50,000 tons/yr of potlining.

In another example, the use of potlining as a fluoride mineralizer has been terminated by American Rockwool for similar regulatory reasons, even though extensive tests demonstrated that the environmental impacts of mineral wool production were reduced with use of potlining. American Rockwool's decision prevented about 7,000 tons/yr of potlining from being reused.

SOURCE: Byers, 1991.

> **BOX 7 Management of By-Product.**
>
> A variety of slags, scales, skimmings, drosses, dusts, sludges, and other by-products result from the manufacture of primary industry products. Some are recycled; some are disposed of in landfills. For example, the production of 1 ton of rolled steel in an integrated plant produces 0.3 ton of blast furnace slag, 0.1 ton of basic oxygen furnace slag, and 0.1 ton of other by-products. By-products containing a relatively high metal content are typically recycled. However, approximately 0.5 ton of by-product material per ton of steel produced is sent to the landfill. The cost for disposal of nonhazardous waste by-product to the landfill ranges from $10/ton to $40/ton; disposal of hazardous by-product materials can cost as much as $300/ton. Identifying uses for these by-products either in PMPI or in other industries throughout the economy would diminish the amount of material disposed of in landfills and offset the significant costs of this disposal.
>
> SOURCE: Carson.

waste in industrial systems (Box 7). There are similar laws in other countries, and there are also international laws, such as the Basel Convention, that prevent residual materials from crossing national borders, where they might be used more productively. The Basel Convention is intended to prevent developed countries from dumping hazardous waste in developing countries, a laudable goal. At the same time, however, the agreement prevents the flow of potentially useful materials that are classified as hazardous waste.

The reclassification of waste is not a simple task. It requires the collaboration of government, industry, and other interested communities. Japan's 1991 Law Promoting the Utilization of Recyclable Resources could serve as a model for waste reclassification (Richards and Fullerton, 1994). This law promotes the idea of the recyclable resource, including the recoverability of energy from waste materials. It establishes target recycling rates for each type of recyclable resource, product priorities for specific industry sectors, and standards for recycled material. The law also recognizes the unique role of PMPI in creating industrial ecosystems that minimize waste in interconnected industries. Through this material classification process, annual releases of solid industrial waste from various sectors can be channeled to basic industries such as cement and steel.

In the United States, regulatory compliance is a complicated, time-consuming process that diverts attention from more productive environmental efforts. The complexity involved in determining how residual material should be handled under RCRA is illustrated in Figure 1. The excessive amount of time spent on compliance-related activity and its complications calls for more transparent regu-

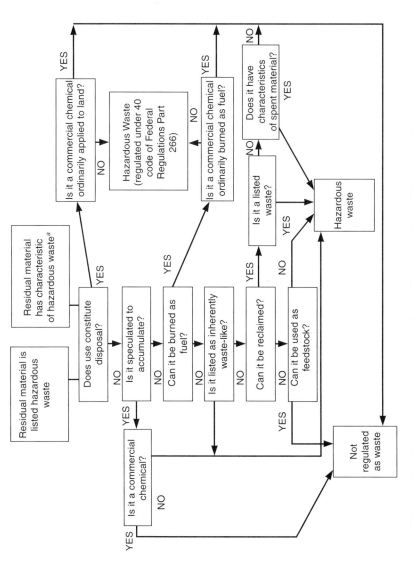

FIGURE 1 A simplified model of hazardous waste material classification. SOURCE: Martchek.
[a]The characteristics of hazardous waste are toxicity, corrosivity, flammability, and ignitability.

lations, better management practices, and better use of information systems within a company. Because regulations have a tendency to become more complicated over time, there is clearly a need for improved methods for handling these information and verification needs.

One outcome of global competition has been the drive to reduce waste, thereby ensuring that as large a percentage as possible of raw materials brought into a company is turned into product. As the cost of complying with regulations escalates, companies are looking for ways to move beyond compliance. In environmental terms, this means minimizing waste and preventing pollution. Once many of the easily identified pollution-prevention measures (such as improving energy efficiency and reducing packaging) are taken, companies look for practical life-cycle approaches to help them in their environmental efforts. Energy-intensive industries such as PMPI recognize that the clean high-temperature burning of waste can reduce energy costs while also benefiting the environment. Ways should be explored to further exploit this potential. At the same time, steps should be taken to devise more sensible regulation and management systems that can streamline compliance with regulations and facilitate pollution-prevention initiatives.

One group working toward improving environmental management is the International Standards Organization's Technical Committee 207, which is developing a set of international environmental management standards known as ISO 14000. It is hoped the standards will help lead to harmonization of such things as rules, labels, use of life-cycle analysis, and environmental auditing. A significant implication of these standards, as with the ISO 9000 series, is that for PMPI facilities to sell their products, they will need to be certified initially in Europe, and then eventually worldwide. In other words, certification could become a de facto requirement for being able to do business in Europe and other regions.

The PMPI sector faces several challenges in dealing with national and international regulations and standards:

- complying with regulations that are at times antithetical to the pollution-prevention approach, which is now hailed as presenting viable solutions to environmental concerns;
- working with regulators to reclassify waste to promote reuse and recycling of materials and energy;
- continuing the search for and implementation of pollution-prevention methods under constrained fiscal and regulatory circumstances; and
- working with international standards-setting organizations to develop reasonable and realistic environmental management systems.

LIFE-CYCLE PRACTICES

Life-cycle assessment (LCA) is an increasingly important aspect of corporate environmental management, particularly for companies trying to reduce the environmental impacts of their products, processes, or services (Box 8). LCA is a tool

BOX 8 Life-Cycle Inventory on 12-Ounce Aluminum Beverage Cans

To evaluate priorities for process improvements and to guide new product development, a comprehensive life-cycle inventory was recently performed by the three major aluminum producers in North America—Alcoa, Alcan, and Reynolds—on 12-ounce aluminum beverage cans.

Although several groups had previously attempted to inventory aluminum products for various applications, those studies lacked information on aluminum manufacturing and recycling processes and up-to-date performance data. Specifically, they did not take into account the aluminum industry's use of hydroelectric power, and they used performance data from the 1970s that did not reflect reductions in process energy consumption, improved gas-collection technology, decreased weight of the product, and increased recycling.

For a life-cycle inventory to be useful to manufacturers, it needs to reflect the most current performance data available. The purpose of this inventory was to provide the three companies with detailed information about materials and energy flows in the life cycle of a beverage can—from the acquisition of the raw material through recycling. The information was used to provide a baseline for improving energy management and the use of raw materials. The study also attempted to highlight areas in which the companies could focus their efforts to reduce the environmental impacts of their operations.

From the inventory, the companies were able to quantify potential environmental benefits associated with three different scenarios: increasing the recycling rate of cans by 6 percent; decreasing the weight of the cans by 2 percent; and reducing secondary packaging (the plastic web that holds the cans together) by 50 percent. (See table below.)

Life-Cycle Inventory of 12-ounce Aluminum Can: Potential Reductions in Energy Use, Air Emissions, Water Effluent, and Solid Waste Relative to Baseline Averages, by Percent

	Best-in-Class Practices	Increase Recyle Rate by 6 Percent	Decrease Can Weight by 2 Percent	Decrease Secondary Packaging by 50 Percent
Energy	10.6	4.2	2.7	11.0
Air emissions	10.2	4.7	2.8	8.9
Water effluent	0.2	0.1	0.1	0.1
Solid waste	23.2	8.0	4.0	9.7

SOURCES: Atkins and Martchek.

that enables the identification of the polluting effects of the energy, materials, and processes used to make products—from the harvesting of the inputs through production, consumption, reuse, and disposal.

Some companies and trade organizations have used LCAs to bolster their green marketing claims. This has led some governments and standards-setting organizations to widen the role for LCA. For example, the European Community is considering a regulation that would make LCA a necessary and sufficient condition for awarding a product a "green seal of approval." In the United States, EPA envisions a prominent role for LCA in designing its own regulatory programs, and several private environmental groups want to use LCA to develop an environmental labeling system for all products.

LCA has several limitations that must be factored into its use for environmental policymaking and regulation. First, every LCA must deal at the outset with daunting boundary questions. For example, in a typical LCA, it would be appropriate to consider the pollution that results from extracting raw material for a product or the quantity of solid waste left behind when a product has reached the end of its useful life. What about the energy and raw materials that went into manufacturing the equipment used to extract the raw materials? Should that be included? What about the capital equipment and labor required to monitor the landfills? In other words, boundary definitions dictate what gets counted in an LCA and what does not. Second, LCAs are expensive, labor intensive, and time consuming. Third, the dynamic nature of technology and changes in production processes make the useful life span of these analyses frustratingly short.

The results of an LCA can be interpreted in a variety of ways and can be used to prove a preconceived hypothesis. Hence, it is important to first define the problem and then identify an appropriate tool to examine the problem objectively. LCA can be effective in aiding pollution prevention, but its robustness as a legislative, policy, or regulatory tool is questionable. In the environmental area, it is more important to develop and adopt a systematic analytical process, which might not be limited to or fully satisfied by LCA.

The use of LCA as an environmental management tool presents PMPI with several challenges:

- addressing the adoption of LCA as a regulatory tool, when methodologies have not been standardized;
- identifying the limitations of LCA when it is used as a pollution-prevention tool so that accurate and appropriate information can be gleaned from such an analysis; and
- working with various stakeholders to develop a standardized LCA methodology.

CULTURAL AND ORGANIZATIONAL CHANGE

PMPI today faces the challenge of responding to society's demands for new, environmentally benign products and processes, and cleaner air, water, and land

> **BOX 9   Alcoa's Use of Reengineering Processes to Improve Environmental Performance**
>
> In 1991, Alcoa's Massena, New York, facility began analyzing its caustic etching operations for many wire, rod, and bar aluminum products as part of a company-wide reengineering program. Caustic etching, which uses caustic soda as well as sulfuric and nitric acid, is used to improve the product's surface quality. The etching process produces various waste liquids, sludges, and rinse waters containing the above-mentioned chemical constituents and metals etched from a variety of aluminum alloys. As a result of the reengineering analysis, the need for the straight-length etching facility that produced relatively simple shapes was eliminated through improved process controls, revised operating procedures, and redesigned die geometry. By December 1994, the entire etching operation was terminated because of process improvements that got rid of the need for the surface-treatment step. Operating costs have been decreased through reductions in the purchase of acids and caustics, elimination of the need to treat and dispose of hazardous solutions and sludges, and enhancement of aluminum recovery from the etching process itself.
>
> SOURCES: Atkins and Martchek.

amid constrained human and financial resources and rapid technological advances. Some companies are reengineering to respond to these constraints and to capture the benefits of new technology (Box 9).

Intense international competition and downsizing have caused many companies to reduce their R&D budgets, including R&D for products and processes that are environmentally preferable. Some have suggested that investment in U.S. national laboratories be used to help offset this reduction in private R&D spending. These laboratories are best known for high-cost, long-range defense- and energy-related R&D. The usefulness of refocusing these efforts to non-defense-related R&D is an important policy question for the United States.

As industry in general grapples with internal pressures, such as reduced R&D capability or reduced human and financial resources, PMPI firms face an additional challenge: their negative image. PMPI companies are seen by many as environmental laggards—smokestack industries belching tons of pollution. This perception has not changed in spite of improved environmental performance by PMPI businesses. PMPI companies have to work hard to change that perception so that they can engage the public in addressing current environmental issues and attract new talent.

Although the face of business has been dramatically transformed by globalization, intense competition, and information technology, the face of government has not changed as dramatically. The anticollaborative nature of government-industry relationships in the United States has forestalled and continues to adversely affect movement toward environmentally sustainable practices.

Perhaps one of the most fruitful approaches for dealing with cultural- and organizational-change questions is to shape tomorrow's engineers and public policymakers. The dramatic changes occurring in industry and the need to develop a more sustainable future have significant implications for academia. In engineering's 4- or 5-year undergraduate curriculum, students acquire the basic scientific and engineering skills to practice their profession. Where do issues such as the generation and control of pollutants, minimization and selection of raw materials, recycling, reduced use of toxic materials, LCA, and the role of public policy fit in this curriculum? What is the role of the professional societies? Although some progress is being made to incorporate these areas of study into undergraduate education, the academic and the professional communities need to pay more attention to environmental education.

Organizational and cultural change is a fact of industrial life. As PMPI companies face the effects of downsizing, globalization, and reengineering, they confront several challenges:

- remaining competitive while adapting to highly volatile business climates that are increasingly influenced by environmental concerns;
- developing mechanisms to leverage external R&D resources to offset downsizing of internal R&D functions;
- identifying ways to use new approaches, such as reengineering, to integrate environmental considerations into organizational changes; and
- working with academia to integrate environmental issues into science and engineering curricula.

REFERENCES

Akimoto, Y. 1994. Materials: Primary resource industries. Pp. 25–26 in Industrial Ecology: U.S.-Japan Perspectives, D.J. Richards and A. Fullerton, eds. Washington, D.C.: National Academy Press.

Aluminum Association (AA). 1993. Aluminum Statistical Review for 1993. Washington, D.C.: AA.

American Iron and Steel Institute (AISI). 1993. Annual Statistics Report. Washington, D.C.: AISI.

Byers, R.L. 1991. Regulatory barriers to pollution prevention. Pollution Prevention Review (2)1: 19–29.

Richards, D., and A. Fullerton, eds. 1994. Industrial Ecology: U.S.-Japan Perspectives. Washington, D.C.: National Academy Press.

Manufacturing

ROBERT A. LAUDISE AND THOMAS E. GRAEDEL

Of all industrial activities, manufacturing is the one that can most easily be environmentally responsible and innovative. Manufacturers are in a unique position. They are not the resource extractor, digging or drilling whatever raw materials have a market; they are not materials processors, forming the powders, crystals, or liquids needed by manufacturers; and they are not marketers, making available to customers whatever goods are produced. While those sectors can exercise some influence, they do not have the freedom of the manufacturer, whose sole constraint is to produce a desirable, salable product. In this regard, the manufacturer can choose to make an automobile body from sheet steel, composites, aluminum, or plastic. Cost, manufacturability, and consumer acceptance are constraints, of course, but the choice of materials per se is not. A telephone transmission system can be coaxial cable, optical fiber, microwave, submarine cable, or satellite. Thus, the designer's role in the manufacturing industry is central, both in the choice of materials and in the choice of process.

Although there are substantial commonalities among the manufacturing sectors, there are great diversities as well, as indicated in Table 1. An important distinction among the sectors is the lifetimes of their products. Some are made to function for a decade or more. Others have lives measured in months or weeks. Still others are used only once. A designer of manufactured goods obviously must adopt different approaches to these different types of products in terms of durability, materials choice, and recyclability.

Regularly in manufacturing, materials and activity trade-offs are evaluated on the basis of market values and regulatory imperatives. Seldom is there a very explicit evaluation based on factors causing an environmental problem. However,

TABLE 1 Manufacturing Sectors and Their Products

Manufacturing Sector	Product Examples	Product Lifetime
Electronics	Computers, cordless telephones, video cameras, television sets, portable sound systems	Long
Vehicles	Automobiles, aircraft, earth movers, snow blowers	Long
Consumer durable goods	Refrigerators, washing machines, furniture, furnaces, water heaters, air conditioners, carpets	Long
Industrial durable goods	Machine tools, motors, fans, air conditioners, conveyer belts, packaging equipment	Long
Durable medical products	Hospital beds, MRI testing equipment, wheelchairs, washable garments	Long
Consumer nondurable goods	Pencils, batteries, costume jewelry, plastic storage containers, toys	Moderate
Clothing	Shoes, belts, polyester blouses, cotton pants	Moderate
Disposable medical products	Thermometers, blood donor equipment, medicines, nonwashable garments	Single use
Disposable consumer products	Antifreeze, paper products, plastic bags, lubricants nonwashable garments	Single use
Food products	Frozen dinners, canned fruit, soft drimks, dry cereal	Single use

in-plant materials and process decisions can be made on the basis of environmental preferability. In addition, a manufacturer may buy its subsystems elsewhere and thereby exert at least a modest influence on upstream environmental activities. Similarly, the manufacturer may choose to perform downstream recycling or at least to enter into cooperative arrangements for doing so. In these ways, the manufacturer can examine and influence trade-offs over the complete product life-cycle, including initial materials choice, product design, processes used to manufacture, in-service impacts, ease of disassembly and reuse, and strategy for recycling and disposal. Thus, even though the manufacturer does not generally have full control, the product life cycle is more solidly under its purview than under that of the processor, the service provider, or the consumer. Therein is the manufacturer's great challenge and opportunity.

ENVIRONMENTAL STEWARDSHIP

The Role of Management

Effective management is essential if manufacturers are to meet the challenge of environmental stewardship. One effort at improving management has been integrated product development (IPD). IPD is aimed at avoiding delays in getting a product to market by incorporating as many external factors as possible in the design stage. As the competitive pressure to get products to market quickly continues to grow, manufacturers are becoming more adept at anticipating problems and designing products to avoid compromising on quality and delays during manufacture, testing, or delivery. Within IPD, a "design-for-X" (DFX) regime has emerged, where X addresses such factors as manufacturability, testability, or maintainability. Design for environment (DFE) is a module within the DFX regime that allows the manufacturer to consider systematically environmental factors, such as concurrent engineering, that can be used in IPD practices.

The DFE module is intended to incorporate traditional concerns about health and safety as well as more contemporary environmental concerns. This approach not only avoids delays that result from overlooking environmental permitting or compliance requirements, but also results in design innovations that meet multiple environmental goals. For example, reducing the mass of a product contributes to resource conservation, since less material and energy are used per unit product, and to health and safety, since less pollution is emitted per unit product. And switching from white bleached-paper packaging to recycled-content packaging avoids the use of chlorine in the bleaching process and can reduce environmental impacts in the large industrial system in which a company operates. Box 1 provides examples of recent innovations in the electronics industry designed to further environmental stewardship.

Stages in the Product Life Cycle

The practice of environmental stewardship is perhaps most conveniently examined against the backdrop of the different stages of the product life cycle (Figure 1). The stages in the product life cycle and associated improvements that can be made at each stage illustrate some of the more common DFE strategies practiced in industry today.

- Stage 1—Acquisition of raw materials, components, and subassemblies. It is a routine practice, though logistically complex, to consider the sources of materials or components and subassemblies from which a product is made to assure product quality. Similarly, with some planning and foresight, the supplier chain can be managed to leverage environmentally preferable inputs and practices in addition to traditional concerns of quality, cost, and performance.

    Purchasers of goods, particularly if they are large customers, such as the government or large service or manufacturing companies, can demand from their suppliers environmentally superior products through procurement documents such as standard-component or stock lists. These documents can be used to specify components that are as environmentally be-

BOX 1   Environmental Stewardship in the Electronics Industry

Recent actions by the electronics industry illustrate the manufacturing approach to environmental stewardship. The electronics sector is neither energy nor materials intensive. Some electronics-based products, such as computers, enable people to avoid traveling, and thus prevent pollution. Even so, in response to the suggestions of environmental groups in Silicon Valley, California, Arizona, and elsewhere, SEMATECH (a semiconductor industry consortium) has been directed by Congress to specifically address environmental issues to minimize toxic waste and emissions. Current industry successes include the following:

- Virtual elimination of chlorofluorocarbons in printed circuit board cleaning, to comply with the Montreal Protocol on substances that deplete the ozone layer. This was made possible through a combination of solder flux minimization, the use of water-soluble fluxes, the use of naturally occurring organic solvents such as limonene, and closed-system cleaning.
- Elimination of glycol ethers in clean rooms. The driving force was evidence of a statistical correlation between increased miscarriages and possible exposure of clean-room workers to glycol ether. The solution was the substitution of ethylethoxypropionate and similar nontoxic compounds as solvents.
- Development of in situ generation of arsine, a highly toxic but industrially vital gas traditionally manufactured elsewhere and transported to electronic manufacturing facilities, with all the potential hazards that accompany the movement, storage, and use of a toxic chemical.

In addition to these successes, there are a number of specific actions that would improve environmental stewardship in the electronics sector. For example, the desirable but yet-to-be-accomplished elimination of silane (SiH_4) and phosphine (PH_3) in processing facilities represent a potential opportunity. Silane is a crucial raw material used in the production of polycrystalline and amorphous silicon. It is also highly pyrophoric and poisonous. Can other, less hazardous sources of silicon be found? Similarly, phosphine is the source of phosphorus in III-V semiconductor epitaxy. It is extremely poisonous, too. Can nonpoisonous phosphorus sources, which are used to produce high-quality light-emitting diodes and lasers, be found? Another area for attention is the very large quantities of water used in integrated circuit wafer processing. Given the likelihood of increasing demands on water resources, water should no longer be considered a negligible factor; rather, its availability and level of use should be scrutinized.

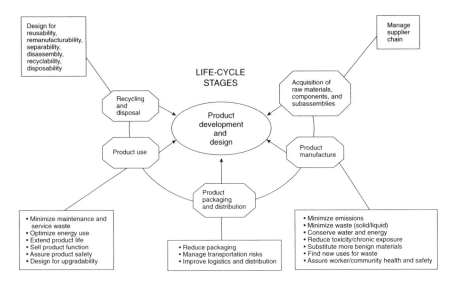

FIGURE 1 Incorporating environmental considerations in development at different stages and of the product life cycle. SOURCE: Richards and Frosch, 1997.

nign as possible. Purchasing contracts can similarly be used to influence suppliers' behavior by requiring environmentally preferable practices.
- Stage 2—Product Manufacture. Environmental strategies in product manufacturing are well-known: minimize air emissions; minimize solid and liquid waste generation; conserve water and energy; reduce toxicity (ensuring worker health and safety during production); and avoid compromising the health and safety of customers, recyclers, and waste handlers. Companies can also seek new uses for waste or ways to convert a waste into a material or energy resource by changing a production process.
- Stage 3—Product Packaging and Transport. Many industries have focused on packaging to improve environmental stewardship. Efforts have included less over-packaging using styrofoam peanuts and bubble packs, elimination of toxic inks from product cartons, reductions in the size of packaging containers, and materials choices designed to enhance the recyclability of packaging.
- Stage 4—Product Use. Environmental options include conserving energy and minimizing waste associated with maintenance and service (particularly for long-lasting products). A company may also want to consider selling product functions instead of product hardware. The company's product then becomes pest control instead of pesticides, for example, or communication instead of phones, computing power instead of computers, refrigeration instead of refrigerators, transportation instead of auto-

mobiles. This idea is not new and it does not apply to all products. It does, however, reflect both a past practice and a growing trend. It was not too long ago that the phone company owned the phones. And the utility industry is now beginning to see itself as a seller of energy-efficient systems rather than kilowatts.
- Stage 5—Recycling and Disposal. Attention to the fate of a product can result in product innovation. Omitting toxic materials from a product would be prudent for a product destined to be discarded. If a product is viewed as inventory to be recovered and reused or refurbished, it may be designed for longer life and ease of upgradability. If the product is to be recycled, it would be desirable to limit the diversity of material. For example, it is preferable to construct the product with one type of plastic rather than several different types. This reduces the amount of separation and sorting needed for recycling. It also helps if the material can be clearly identified for recyclability, such as by molding identification marks into plastic parts. These are examples of strategies that are emerging as the environmental life-cycle approach is applied to product design.

If products are to be returned to the manufacturer, their recovery and associated logistics would have to be managed in much the same way that supplier chains are managed. Type, size, and life of product influence the needs of the recycling infrastructure. In the United States, for example, a recycling infrastructure has developed that recovers 75 percent, by weight, of the materials in automobiles (Klamisch, 1994). Creative solutions are also being developed to address the question of what to do with the electrophotographic cartridges used in laser printers and fax machines.

Although the cartridges are used in large volumes, and some larger customers provide a single point for recovery, many customers are small-scale users. The challenge has been to amass sufficient volumes of cartridges for economical and efficient recycling. In the United States, direct collection from users through prepaid parcel services has been used, while in Europe dealers serve as collection points. Collection, however, is just part of the equation. Companies that sell these electrophotographic cartridges have also invested in recycling centers and new recycling and remanufacturing processes.

Life-cycle stage 2 and, perhaps, life-cycle stage 3 have been regarded customarily as falling within industry's responsibility, but the evolving view is that an environmentally responsible product minimizes external environmental impacts in all five life cycle stages. To date, few manufacturers have looked at the whole life cycle of their products and asked questions such as, Do our products use abundant resources rather than scarce ones?, Are our products made from recycled rather than virgin materials?, Do our products minimize energy use?, and Are our products designed to be efficiently disassembled at obsolescence? The manufacturing sector has traditionally computed the energy efficiency of its processes (though substantial room for gains in efficiency remains). The sector must now begin to consider the materials efficiency of processes, with the goal that every molecule entering a manufacturing facility goes where it is needed to perform a useful function.

The movement toward environmental stewardship could be extended from products and processes to manufacturing facilities. All such facilities might be expected ultimately to consider the environmental impacts of the site during site selection and development; participate in industrial infrastructures, sharing excess heat and process by-products; and operate with a focus on environmental stewardship in such areas as heating, lighting, property maintenance, and waste disposal.

EFFECT OF REGULATIONS AND STANDARDS

"End-of-Pipe" Laws

Much current environmental law, as it affects manufacturing, reflects the perception that environmental problems as localized in time, space, and media (i.e., air, water, soil). For example, it was not uncommon in the recent past for organic-solvent of groundwater contamination to be eliminated by "air stripping," or simply releasing the solvent to the air, where it contributed in many cases to the formation of tropospheric ozone. Similarly, many hazardous waste sites in the United States have been "cleaned" by shipping the contaminated dirt somewhere else, which not only does not solve the problem, but also creates the danger of incidents during removal and transportation.

As a consequence of this perception, environmental regulations have focused on specific phenomena and adopted a "command-and-control" approach, in which restrictive and highly specific legislation and regulations are implemented by centralized authorities and used to achieve narrowly defined ends. Such regulations generally prescribe very rigid standards, often mandate the use of specific emission control technologies, and generally define compliance in terms of "end-of-pipe" emissions limits. Examples in the United States include the Clean Water Act (by EPA interpretation applied only to surface waters), the Clean Air Act, and the Comprehensive Environmental Response, Compensation and Liability Act, or "Superfund."

Many environmental laws are based on assumptions appropriate to a linear flow of materials to waste, rather than the internal cycling characteristic of a sustainable economy. For example, the Resource Conservation and Recovery Act defines almost any by-product of a linear manufacturing process as a hazardous waste and subjects it to burdensome regulation, thereby limiting the incentives to recycle or reuse the material. The regulation thereby tends to institutionalize the linear manufacturing paradigm, rather than guide industry toward sustainability.

If properly implemented, command-and-control can nonetheless be effective in addressing specific environmental insults. For example, U.S. rivers such as the

Potomac and Hudson are much cleaner as a result of the Clean Water Act. Moreover, where applied against particular substances, such as the ban on tetraethyl lead in gasoline in the United States, the command-and-control approach has clearly worked well. Such regulations, however, are characterized by a burgeoning of mandatory requirements, a relative lack of concern for economic efficiency, a focus only on the manufacturing stage of industrial activity rather than the life cycle of materials or products, and, because specific technologies are prescribed, a strong bias against technological innovation. Moreover, in practice, it has proved very difficult to modify such regulations to reflect advances in scientific understanding.

Activities of International Standards-Setting Organizations

Given its record of formulating international standards, the International Standards Organization (ISO) was a natural place to begin the process of generating standards for sustainable industrial development. Accordingly, 1991 saw ISO establish in coordination with the Geneva-based Business Council for Sustainable Development, its Strategic Advisory Group on the Environment (SAGE). SAGE subgroups focus on life-cycle assessment (LCA), environmental guidance for product standards, environmental management and auditing, and environmental labeling. Over the next several years, standards will begin to emerge from these efforts, probably along the lines of the ISO 9000 Quality Standards and Methods program.

Another international organization with an interest in environmental stewardship is the International Electrotechnical Commission (IEC), an independent body that has a long record of generating standards and protocols to assist in the globalization of technology. Focusing on international telecommunications technology, particularly electronic equipment, an expert group of the IEC has been concerned with several issues related to environmental standards:

- the primitive state of LCA;
- pollution prevention, particularly actions to eliminate solder cleaning and minimize the use of deionized water;
- environmental impact assessments, especially of ozone-depleting substances, batteries, consumables, packaging, and emissions; and
- design for disassembly, with attention to fastening and joining and on decorative paints and finishes.

The main question with international environmental performance standards is not whether they will be put in place, but how soon. Farsighted corporations would be well advised to prepare so that the imposition of standards will place them in a favorable position with respect to their competitors; and participate in standards-setting activities, so that the standards that do emerge are a reasonable consensus of the positions and desires of all concerned.

Waste Minimization, Packaging, and Product Take-Back

In many countries and among many customers, corporate environmental stewardship is receiving increasing attention, but the focus is turning toward setting goals for performance rather than regulating specific actions. In parts of Europe where this effort has been most vigorously pursued, complete packaging take-back is required. A customer buying a tube of toothpaste in Germany, for example, can immediately hand the empty box back to the merchant and, if he or she wishes, return in a month to hand back the empty tube. Packages are becoming smaller, are featuring less diversity of materials, and are being more efficiently recycled.

The next step, and one likely to occur first in Europe, is the requirement that manufacturers take back their products if they become obsolete or need replacement. Products that have been designed with take-back in mind will at that time be efficiently disassembled and their components and materials readily reused. Products not so designed will, in all probability, be landfilled at substantial cost.

Labeling Programs

A number of labeling systems designating environmentally responsible products and/or activities are being developed around the world. Some of these systems are designed by marketers whose goal it is to advertise the environmentally beneficial characteristics of individual products or of the corporation itself. While these "first-party" systems are useful in demonstrating an environmental focus, many of them are viewed by experts as self-serving and potentially inaccurate. These activities may nonetheless provide stimulus either for customers or corporations to change their purchasing decisions.

More visible and more effective are environmental labeling programs. Some are quasigovernmental; others are strictly private. In principle, seal-of-approval labels (Figure 2) are awarded as a result of some sort of life-cycle assessment. (See section on LCA, p. 55.) The standards tend to be set so that a modest fraction of existing products, perhaps 10 or 20 percent, can successfully qualify. Since LCA in its full embodiment is complex and contentious, labeling organizations tend to use a simplified LCA, especially as far as the impact analysis stage is concerned. In practice, one life-cycle stage or one product characteristic (toxic content, say, or diversity and volume of packaging materials) often controls the labeling decision, so labels seldom reflect the overall environmental impact of a product.

In many cases, the criteria for obtaining environmental labels for specific products are far more stringent than existing regulations or standards. Nonetheless, corporations are often driven by competitive pressures to satisfy the labeling requirements. The advantage of labeling systems is that they harness market forces and consumer preferences to achieve better environmental performance. At a stage of industrial ecology when life-cycle assessment methods are still relatively unfamiliar, labeling programs offer the promise of rapidly accelerating the use of

FIGURE 2 Some of the labels established to designate environmentally responsible products; (a) German "Blauer Engel"; (b) Nordic Council "White Swan"; (c) Canadian "Environmental Choice"; (d) Japanese "Ecomark"; (e) Dutch "Stichting Milieukeur"; (f) United States "Green Seal"; (g) Singapore "Green Label"; (h) European Community "Ecolabel"; and (i) United States "Energy Star." SOURCE: Graedel and Allenby, 1995.

LCA methodologies. However, a potential disadvantage of labeling programs is that unless their criteria are carefully chosen, environmentally suboptimal performance might be encouraged.

ENVIRONMENTAL METRICS

As companies report on their environmental successes, they face external pressures to prove these claims. A range of environmental metrics are emerging for both external reporting purposes and internal management needs (Torrens and Yeager, this volume). The simpler metrics relate energy, mass, and volume measured in absolute terms or normalized with respect to production unit. These measures can be further refined to provide better information. For example, total energy use in manufacturing can be broken down in terms of renewable and nonrenewable sources; total material use can be broken down into toxic or nonhazardous material used; and total industrial waste generated can be broken down to get risk exposure information for ambient concentrations of hazardous waste by-products in water, air, or soil. As reuse and recovery take hold, other measures are being used, such as time taken for disassembly and recovery; percentage of material or product reused, recycled, or disposed; and percentage of recycled material used as inputs. Useful operating life can also be translated as an environmental measure. Businesses respond most strongly to profits and cost savings, and so translating savings associated with design improvements can be a strong motivator for further change.

More difficult are the measures needed to compare the consequences of selecting among functionally equivalent materials and processes. "Environmental preferability" depends on the boundary conditions of the analysis as well as on value judgments. Functional equivalency is more complicated when attempts are made to balance costs and benefits. For example, how do the environmental consequences of manufacturing electronic switches and computers rate compared with those of using switches and computers in systems that facilitate telecommuting and thus reduce substantially environmental impacts of travel?

LIFE-CYCLE ASSESSMENT

An LCA involves doing an inventory of materials and energy inputs and outputs, analyzing the impacts of inputs and outputs, and prioritizing actions that may be taken to address these impacts. In practice, it has been difficult for corporations to carry out detailed life-cycle inventories, more difficult to relate those inventories to a defensible impact analysis, and still more difficult to translate the results of those LCA stages into appropriate actions. The principal problem has been that comprehensive life-cycle inventories are expensive and time consuming. Another difficulty is that impact analyses connected with an assessment are inevitably contentious, and numerical assignments of impact are not accepted as adequate guidance. Finally, it is hard to rank one new product design against another or an old product against a new product.

Although it is easy for managers to support the principle of integrating environmental factors into decision making, committing the organization to doing so as a matter of course requires developing a standard approach and a standard measuring system. Experience seems to demonstrate that the process works best when it is purposely done in modest depth and in a semiquantitative manner. The goal is to do the LCA rapidly, perhaps within 2 days for a typical product and within 1 week for a typical facility. One LCA method has as its central feature a five-by-five Environmentally Responsible Product Assessment Matrix, one dimension of which is life-cycle stage and the other of which is environmental concern (Figure 3). Using this matrix, the design-for-environment assessor studies

Life-Cycle Stage	Environmental Concern				
	Materials Choice	Energy Use	Solid Residues	Liquid Residues	Gaseous Residues
Resource extraction					
Product manufacture					
Product packaging and transport					
Product use					
Recycling, disposal					

FIGURE 3 The Environmentally Responsible Product Assessment Matrix. Considerations and desirable actions pertaining to each matrix element are described by checklists and recommendations specific to the type of product under evaluation.

the product design, manufacture, packaging, in-use environment, and likely disposal scenario and assigns to each matrix element a rating from 0 (highest impact, a very low evaluation) to 4 (lowest impact, an exemplary evaluation). In essence, what the assessor is doing is providing a numerical figure of merit to represent the estimated result of the more formal inventory analysis and impact analysis. The assessor is guided in this task by experience, a design and manufacturing survey, appropriate checklists, and other information.

In arriving at an individual matrix element assessment, or in offering advice to designers seeking to improve the rating of a particular matrix element, the assessor can refer for guidance to underlying guidelines, checklists, and protocols. Examples of questions to guide scoring are given in Box 2 for three of the matrix elements.

Once an evaluation has been made for each matrix element, M_{ij}, the overall environmentally responsible product rating (R_{ERP}), can be computed as the sum of the matrix element values:

$$R_{ERP} = \sum_{i=1}^{5}\sum_{j=1}^{5} M_{ij}$$

Since there are 25 matrix elements, a maximum product rating is 100. Several trial uses at AT&T of this scoring system have shown that its demands on time and resources are modest and its results readily communicable.

A number of companies are beginning to use such matrices. For example, Volvo uses an environmental priority system (Box 3) that provides a numerical ranking of various materials (Horkeby, 1997). AT&T uses the matrix method described above to evaluate environmental characteristics of interest across life stages of products (Box 4). In both cases, complex decisions are involved in

BOX 2 Sample Checklist Questions for Elements in the Environmentally Responsible Product Assessment Matrix

For the semiquantitative evaluation of the matrix in Figure 3, each matrix element includes checklists and recommendations. Some of the items appearing there will be common across all manufacturing sectors, while others will be specific to a particular sector. In order to provide perspective on this process, a few samples of checklist questions for three of the matrix elements are given below.

Matrix Element 1,1
Life-Cycle Stage: Resource Extraction
Environmental Concern: Materials Choice

- Are all materials the least toxic and most environmentally preferable for the function to be performed?
- Is the product designed to minimize the use of materials in restricted supply?
- Is the product designed to use recycled materials wherever possible?

Matrix Element 2,4
Life-Cycle Stage: Product Manufacture
Environmental Concern: Liquid Waste

- If solvents or oils are used in any manufacturing process in connection with this assessment, is their use minimized and have substitutes been investigated?
- Are liquid product residues designed for minimum toxicity and optimal reuse?
- Have the processes been designed to use the maximum amount of recycled liquid species from outside suppliers rather than virgin materials?

Matrix Element 5,3
Life-Cycle Stage: Recycling, Disposal
Environmental Concern: Solid Waste

- Has the product been assembled with fasteners such as clips or hook-and-loop attachments rather than chemical bonds or welds?
- Have efforts been made to avoid joining dissimilar materials together in ways difficult to reverse?
- Are all plastic components identified by ISO markings as to their content?

assigning values to either an "environmental load unit" (as in the Volvo system) or in judging the extent of environmental impacts (in the AT&T matrix analysis). They are, however, useful compasses for guiding improvement.

CULTURAL AND ORGANIZATIONAL CHANGE

Cultural and organizational changes are critical to the success of corporate operational responses to environmental issues. Many of these changes are within

---

BOX 3 Volvo's Environmental Priority Strategies System

Volvo's environmental priority strategies (EPS) system translates environmental data into a format that can be used by design teams. Volvo developed the EPS system in collaboration with experts from academia and other companies. The firm owned or controlled much of the necessary data, such as the energy costs and environmental impacts associated with making a kilogram of steel. This information was obtained from the Federation of Swedish Industries covering a broader range of industrial activity.

The EPS system is based on environmental indices calculated for specific materials. Each environmental index takes into account

- Scope (general impression of the environment impact),
- Distribution (extent of affected area),
- Frequency or Intensity (regularity and intensity of the problem in the affected area),
- Durability (permanency of the effect),
- Contribution (significance of 1 kg of the emission of the substance in relation to the total effect), and
- Remediability (relative cost to reduce the emission by 1 kg).

Environmental Index = Scope x Distribution x Frequency or Intensity x Durability x Contribution x Remediability

The environmental load unit (ELU) per kilogram of any substance is then calculated by multiplying the environmental index by the amount of substance released to the environment. Given the uncertainties inherent in many environmental data and analyses, another necessary aspect of the EPS methodology is the ability to perform sensitivity analyses (i.e., determine what data or change in environmental impacts would change the value of the ELU).

The EPS system is not perfect. It is, however, a tool that company engineers can use to compare the environmental impacts of new products or processes, and as such, it should lead to improved decision making (Horkeby, 1993).

BOX 4 Qualitative Matrix Analysis of Substitutes for Lead Solder

Three alternatives—indium alloys, bismuth alloys, and isotropic, conductive epoxy systems (using silver as the filler)—to using lead solder in electronics were evaluated to rank the environmental concerns associated with each. The alternatives were qualitatively analyzed using the matrix system shown here.

The results of the analysis were counterintuitive. The expected result, that significant substitution of both indium and bismuth alloys would be preferable to the current use of lead-based solders, was not supported. Lead solders proved to be preferable, not because of their obvious manufacturing advantages, but because of their more moderate environmental impacts. These environmental impacts did not occur either at the electronics assembly stage or during consumer use, but rather at the time the materials were mined and the metals were processed.

The environmental impacts associated with indium and bismuth were large because these elements, and to a lesser degree silver, generally appear only in relatively low concentrations in ores. Moreover, these materials are by-products of producing other metals (principally lead, copper, and zinc). So, the environmental impact of the alternatives includes the energy and environmental effects of mining and processing of the primary ores. These effects can be significant in relation to those associated with lead.

In addition, natural deposits of indium, bismuth, and to a lesser extent silver are much smaller than those of lead. This has implications both for cost and for absolute availability. For example, it would require 11,200 metric tons of indium to substitute for the lead used in solder, but world recoverable reserves as of 1991 were only 1,692 metric tons, and the world reserve base is only 3,012 metric tons. Thus, there is not enough indium or bismuth in the world to substitute for any significant portion of the existing lead solder used in printed-wiring board assembly.

Further, the available data do not demonstrate unequivocally that the substitutes are much less toxic than lead. Bismuth probably is less toxic—after all, it is the active ingredient in Pepto-Bismol.® Limited data on the other two substitutes indicate both may be toxic. The matrix system provides a way of displaying the level of uncertainty associated with particular measures.

This analysis indicates that the complete substitution of indium, bismuth, or silver solders for lead solder in printed-wiring board assembly is questionable from both an environmental and a social perspective. Other approaches, including lead recycling and recovery, and connective systems using little if any metal, should be vigorously explored.

SOURCE: Allenby, 1994

(Please see figure, pp. 60–61.)

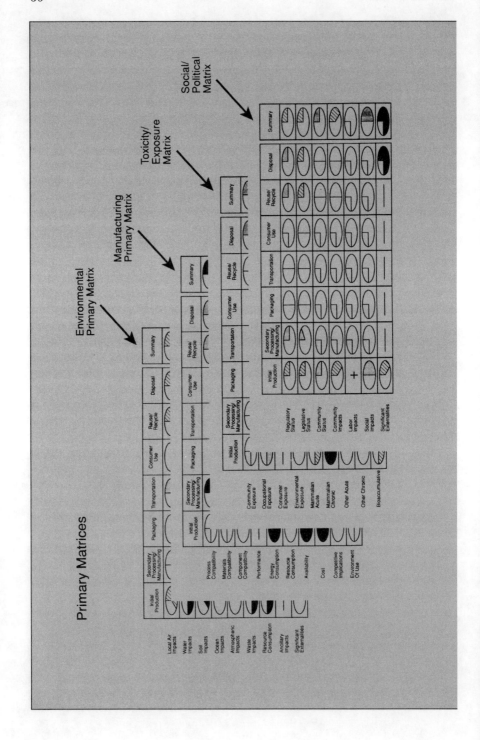

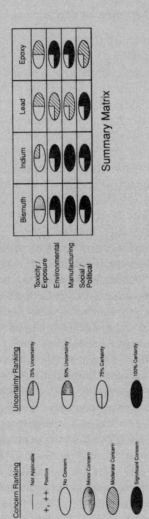

Box Figure 4   Qualitative matrix anlysis of substitutes for lead solder. SOURCE: Allenby, 1994.

the purview of manufacturers; others have much more to do with society at large. This paper examines three aspects of this change: managing environmental costs, altering the structure of demand, and recognizing the key role of the designer.

Managing Environmental Costs

Developing acceptable methods to guide product and process designers in selecting low-environmental-impact materials and production methodologies is not a trivial matter. If environmental costs were reflected in the market price of goods and services, environmental preferability would be apparent. However, environmental costs are often ignored in the national accounting system and are at best hidden as overhead in corporate accounting practices.

The financial characteristics of any manufacturing activity are understandably of interest to industrial managers, and such information is generally captured in management accounting systems. Traditionally, such systems have treated environmental costs—even real, quantifiable environmental costs, such as those associated with residue disposal as overhead and have therefore not broken them out by activity, product, process, material, or technology. Without access to environmental cost information, managers have had neither the incentive nor the data to reduce costs such. The solution, sometimes called "green accounting," is conceptually simple: Develop management accounting systems that break out such costs, assign the costs to the causative activity, and then rationally manage them. One example of such an approach is activity-based costing (ABC) (Macve, 1997; Todd, forthcoming).

In practice, however, green accounting has proved difficult. For one thing, managers tend to resist taking on additional responsibilities. A bigger hurdle in many complex manufacturing operations, however, is developing sensors and systems to provide data on the contributions of different processes and products to a liquid residue stream. In addition, firms may resist assigning potential costs, such future regulatory liability for current residue disposal practices because of the fear of legal liability. (A company that considers potential future liability may be seen as admitting its planned behavior was inappropriate or illegal.) Nonetheless, it is clear that the addition of green elements to management accounting systems and their supporting information subsystems is critical to completing a necessary feedback loop for environmentally appropriate corporate behavior. Continued research into so-called full-cost accounting by the accounting community would be most useful to industrial ecologists.

Altering the Structure of Demand

A different perspective on environmental stewardship occurs once ownership of a product is conveyed to a customer. Decisions concerning environmentally beneficial actions tend to differ according to whether the decision makers are individuals or a group. Individual owners of automobiles, for example, have less ability to influence the environmental fate of the vehicle they own than owners of electricity generating facilities have influence on the environmental practices and operation of their power plants. It is obviously easier to undertake voluntary or mandatory corrective action if ownership and decision-making are centralized than if ownership and decision-making are widely distributed. Thus, chances for improving the environment could be enhanced if trajectories of development could be designed to concentrate sources of environmental impacts in the hands of fewer decision makers and in fewer locations. Doing so involves what social scientists term "altering the structure of demand."

In the last few years of the twentieth century, substantial change is anticipated in the structure of demand as a result of corporate and political actions. The result will be to place the ownership of many goods, such as automobiles and refrigerators, in fewer hands. The goods will then be leased to individual customers and businesses. Once the structure of demand is changed, many factors within and outside the restricted-ownership circle will produce actions leading to fewer environmental impacts. This change turns out to be natural once human preferences are considered. Customers do not buy 1,000 kilograms of metal, 100 kilograms of plastic, and an assortment of mixed materials because they have an innate desire to own them; they buy a bundle of functionalities manifested in the form of an automobile.

One interesting effect of this anticipated transformation is that companies that used to think of themselves as manufacturers—providers of things—will become service companies. To stick with the automobile example, it seems likely that vehicles will increasingly be leased rather than owned. Customers will pick up a leased vehicle, use it as they need to, and return it to the automobile leasing center. The leasing agency, meanwhile, will be responsible for the flow of materials into the vehicle, its manufacture, all maintenance and life-extension activities, and the eventual dismantling of the automobile for appropriate recycling of subassemblies, components, and materials.

A shift in corporate focus to selling function instead of hardware will require organizational and information systems that offer superior inventory or fleet operation. It will require designing systems, products, and components that require minimum maintenance. Such strategies might include developing modular designs that accommodate changes in technologies or user requirements through upgrades. Finally, the shift will also require managing risk and consumer satisfaction at all levels within the complex manufacturing, delivery, use, and disposal systems of which the product is a component. It may prove to be more economical and effective to improve system performance than product performance.

The transition from sales to service will be an enormous culture shock for many manufacturing companies. It will involve the addition of service-oriented marketing practices to more traditional technologically driven corporate practices. The conglomerates of the twenty-first century will be built upon the remains of corporations unable to manage the transition.

Recognizing the Key Role of the Designer

In order to effect major progress in the environmental stewardship of the manufacturing sector, corporate culture will need to be realigned so that the designer of products and processes is both the target of change and the beneficiary of corporate enlightenment. It is not unrealistic to regard the designer as the nexus of environmental decisions in the industrial corporation. Designers choose the systems and subsystems needed to perform the required functions, the components from which to manufacture the systems, and the materials and manufacturing processes.

In the past, the constraints on the designer have generally been to achieve the required function while minimizing cost. Engineering education, handbooks for designers, professional support systems, and corporate reward structures have reinforced these constraints. Environmental concerns have typically been addressed only if legally necessary. The industrial ecology life-cycle ethic will demand of the designer a different educational background and new and expanded databases and tools. Concurrently, new corporate support and reward systems will be required. A central focus of environmental and engineering education, law and regulatory practice, and corporate management practices will be to provide these necessary new tools and incentives for the product and process designer, whose choices will determine to a great degree the environmental performance of the corporation.

BREAKING BARRIERS AND CREATING OPPORTUNITIES

Critical barriers to lessening the impact of manufacturing on the environment have to be addressed at least on two levels: those that fall within manufacturing organizations themselves and external societal and institutional factors that affect the manufacturer. The line between the two is blury, as the following discussion illustrates.

Intraorganizational Barriers

1) *A commitment by management to environmental stewardship is not always evident.* One way a company can demonstrate its responsibility for the effects its products and processes have on the environment is to articulate clearly its policy to the public and to make sure that the policy is backed by a plan to achieve specific goals. Demonstration of a commitment to environmental goals can be further strengthened through membership in appropriate industry organizations that have well-developed principles of self-regulation and oversight. The commitment to these principles must come from the most senior levels of the firms; in most instances, the CEO signs an agreement to abide by the principles and commits to disclose publicly the results of annual reviews by peer member companies.

2) *A focus on financial performance may lead to inappropriate short-term environmental decisions.* Metrics and reward systems must be revised to support the organization's commitment to make manufacturing decisions in light of their impact on the environment. These metrics and reward systems should encourage rational examination of alternative decisions that may result in lower financial performance in the short term but higher environmental and economic rewards over the long term.

Improved methods of manufacturing cost accounting are available to support these decision processes. Activity-based costing is one methodology that can be used to help identify and collect the appropriate data by which alternatives can be evaluated.

3) *Insufficient information is available for making environmentally responsible decisions.* To address this deficiency, handbooks and literature can be assembled to provide the information needed by product designers, engineers, and the managers of manufacturing firms. This collection of guidelines and data must reveal costs throughout all stages of the product life cycle. Designers, engineers, and managers require a rational basis on which to select materials; consider alternative product configurations; compare different production processes and packaging schemes; understand energy use at each stage of the product's life; and consider remanufacture, disposal, or recycling costs at the end of the product's useful life. In larger organizations, the environmental health and safety office could provide this information, but the same issues must be addressed by smaller firms as well. Again, better accounting methods for allocating costs are available to ensure more environmentally responsible decisions.

4) *Externalizing costs has been an acceptable way for manufacturers to transfer to the customer the responsibility (i.e., costs) for a product's environmental impact.* Changes in accounting standards and methodologies are needed before most organizations are able to determine the costs associated with internalizing what are currently treated as external costs. Activity-based costing can help, but it is not as simple as its acronym suggests. ABC increases the volume of data collected and may require accompanying organizational changes and adjustments to accommodate those needs.

While accounting practices can support the operational aspects of internalizing external costs, the encouragement for doing so is unlikely to come from the manufacturing enterprise. Motivation for these changes is more likely to derive from regulatory actions. One option is to place taxes on materials (e.g., mercury) that are of particular concern.

Legislative models that may be more effective at internalizing previously external costs should be developed and tested. Government, industry, and the public need to work collaboratively to identify opportunities to minimize the impact of manufactured products on the environment by examining the influence of the product at each stage of its life. However, these analytical life cycle models can be used inappropriately. LCA can provide useful information for planning purposes within industry, but methodological problems and uncertainties make LCA entirely inappropriate as a regulatory mechanism.

Take-back laws, which require the manufacturer to recover the product from the customer at the end of its life, are also worth examination. But it is not clear what impact this change in responsibility will have on the upstream suppliers of components and subassemblies, or manufacturers will communicate their new approach to the public (e.g., manufacturers of office chairs in Germany advertise that they "want their products back").

Societal and Institutional Barriers

1) Regulatory response often occurs too late to prevent environmental damage. Institutional response can lag behind the initial environmental damage caused by economic, technological, demographic, or behavioral pressures by 10 to 20 years (Rejeski, 1997). One solution to this dilemma is for government agencies and industry-sector consortia to work to anticipate the consequences of emerging technologies. For instance, by the year 2010, about 52 million batteries will have to be recycled, disposed of, or remanufactured as a result of the California electric vehicle initiative. Now is the time to examine the battery technology to be used, the methods of manufacture, and the eventual end-of-life issues for the power sources.

Another example relates to the "green" lightbulb, which was rushed into production by U.S. firms in the 1980s without due consideration for the eventual disposal of the product. The mercury content of the lightbulb was found to cause significant landfill problems, and manufacturers who believed they were producing an environmentally responsible product were ultimately fined. By contrast, a predictive, preventive approach was taken by the Dutch. They investigated the mercury content of the lightbulbs and recognized a possible disposal problem. They developed a bulb with a lower mercury content before commercializing the product and put in place a reclamation structure for recapturing used bulbs.

The goal of approaches like this is to examine a product's "toxic fingerprint" before it is placed on the market. It requires a strong scientific base for environmental decisions and an understanding of technological innovation. The early-examination approach introduces a barrier of its own, however. Companies may be reluctant to participate because there is no assurance that the regulatory environment will not change, and if it does, information about their new development of products (on a 15-year time scale) could be used against them in the future. Still, participation in such efforts can help a company avoid investing in products that are headed for the "endangered" list.

2) Businesses too readily acquiesce to accusations of environmental irresponsibility. Manufacturers and their industry representatives must play a stronger role in informing and educating the public about the environmental consequences of their products, manufacturing processes, and distribution methods if they expect to remain competitive. They must work to move public discussion of environmental issues away from confrontation toward consideration of trade-offs among desired but conflicting objectives.

3) *The evaluation of environmental risks in the context of the entire product life cycle is not well understood by the public.* Significant efforts have been devoted to the development of educational materials, based on case studies and data bases, that introduce life-cycle-based assessment of environmental risk into K-12 and college curricula. Better outreach is needed to advise teachers of the availability of these resources and to help them include a balanced presentation of the issues involved.

4) *The public mistrusts business because in many instances companies have not taken responsibility for problems others felt they were liable for or have presented conflicting information about the "greenness" of their products.* Businesses need to be open with the community in which they are located about their environmental efforts. While there are national and international industry organizations that can provide educational materials and assistance, direct contact with local business leaders is the most effective way for manufacturers to establish their credibility and make clear their contribution to economic development and environmental quality. Manufacturers can also engage the public in discourse about the relationship between industry and the community.

5) *Confrontational relationships exist among business, regulatory agencies, and environmental organizations.* There are several ways to promote anticipatory, nonconfrontational problem solving among all interested parties. These include improving the credentials and salaries of government officials; developing lists of environmental action items; and providing incentives to encourage collaborative problem solving. Upgrading the credentials and compensation of agency officials will be a slow and difficult process. In the meantime, industry sponsorship of travel and registration costs could encourage these officials to participate more in meetings, conferences, and other forums. These meetings would provide opportunities for the joint identification of issues whose solution would provide substantial benefits when cooperatively undertaken.

Immunity from future regulatory action is an important requisite for companies that agree to participate in reducing confrontational relationships with regulators and other concerned groups. Such a provision will help to ensure that companies are forthcoming and engage in frank conversations about problems and suggestions for solutions.

6) *The most effective means of sharing responsibility for environmental stewardship between communities and manufacturers is not articulated.* Since the

responsibility for the environment does not lie solely with industry but requires broad-based involvement of the community, it is useful to build linkages for cooperative action among industry, the community, elected officials, and environmental organizations. A systems view could be encouraged, such as studying whether optimized traffic control is more important for the local environment than a certain type of emission control on an industry.

7) *The increasing influence of national regulatory actions and proposed international metrics and standards could adversely affect the manufacturing community.* The many existing environmental standards and regulations around the world adds complexity to the task of environmental management for globally competitive companies. This will likely remain a concern into the future. At the same time, international efforts to develop standards for environmental management (i.e., the International Standards Organization's ISO 14000 program) are running into conflict with local and regional voluntary schemes. For example, the ISO 14001 standard for environmental management systems is sufficiently different from the European Union's voluntary ecomanagement and auditing scheme (EMAS) that the European standards body has plans to draw up a document to bridge differences between ISO 14001 and EMAS.

If successful, these efforts will document public outreach practices that encourage greater environmental responsibility by firms. The benefit to assembling industry-sector-specific as well as generic best-in-class information is that the data can be used in a manner similar to the algorithms developed for quality improvements.

UNRESOLVED QUESTIONS

The manufacturing sector is currently in a period of transition as it wrestles with whether (and how) to close material loops and create a more sustainable industrial ecology. As this process unfolds, it may be useful for those involved to ponder the following questions:

- How accurate and detailed must life-cycle assessments be in order to support the crucial life-cycle costing activity?
- Under what conditions should external environmental costs be internalized?
- What can be learned from past successes and failures?
- What opportunities exist to harvest the "low-hanging fruit" (i.e., simple and inexpensive activities yielding substantial benefits) of environmental improvement in the manufacturing sector?
- Can the manufacturing industry really know what actions constitute environmental stewardship without close collaboration with environmental scientists?

- Is it realistic to expect the creation of global standards for environmental performance?
- Can manufacturers truly expect to become suppliers primarily of leased items (i.e., providers of function) rather than of products for sale? What roadblocks stand in the way of such a transformation?
- Can groups with a vested interest in maintaining the status quo be convinced to change? Individuals with corporate environmental compliance and control responsibilities may feel threatened by initiatives, such as pollution prevention, that may eliminate the need for their positions.
- Is there real value in green labeling? The value that the public places on green labels depends on whether they trust the source of information. There is currently no method for regulating such labeling programs. In the United States, green labeling is a private-sector activity undertaken by a handful of consulting companies. In Europe, the government role is stronger, and some standards have emerged. However the value of the label is also quickly lost if the green attribute becomes the industry standard. For example, the U.S. EPA Energy Star program to label computers that power down when not in use probably aided the introduction of the innovation across the computer industry. This adoption diminished the value of the label as a differentiator among similar products, although the consequences of not carrying the label remain unclear.
- Is LCA the ultimate decision tool to aid in determining environmental preferability? LCA is incredibly data intensive and value laden, and it lends itself to differing interpretations (e.g., the disposable diaper industry may find its diapers have environmental advantages over cloth diapers, while the marketers of cloth diapers can argue effectively that cloth is superior). Nevertheless, companies are using scaled-down assessments to guide their decision making.
- Can general DFE-based tools (including training) for product and process designers be developed? Large companies such as AT&T and Volvo have begun developing their own tools, but there is a dearth of tools for use by small- to medium-sized companies.
- How can the adversarial relationship between industry and environmental regulatory agencies, which inhibits change, be overcome? A critical element of the cultural and organizational change needed involves a redefinition of the manufacturing industry as one that is environmentally conscious. This will require industry to partnership with customers and government. But such efforts are inhibited by the proliferation of regulatory initiatives. (There were fewer than 50 pages of EPA regulations in 1969, but there are expected to be 11,000 pages by 2010.) In addition, the regulatory system in the United States is one based on a command-and-control approach, which perpetuates a vicious cycle: Government develops the most stringent regulations possible to address a problem (so it can deal

from a strong negotiating position); industry balks and engages expensive lobbyists and lawyers to negotiate its terms while at the same time claiming the action will result in the loss of jobs and undermine competitiveness; industry finds a way to meet the negotiated/lobbied regulations; government follows with ever-more stringent rules; and the cycle is repeated. Throughout the process, time and valuable human resources are expended. More effective partnering is needed to move away from this process toward a flexible systems-based problem-solving mode.

- How can society avoid national regulatory actions and those imposed by international agreements that have a negative impact on innovation and the ability to compete? Market-based regulations have in some instances encouraged innovation (as in the case of finding replacements to ozone-depleting substances). In most instances, however, companies burdened by additional rules and procedures are unable to respond as quickly to changes in markets because of the time and resources consumed with permit and regulatory compliance. Companies from countries not imposing the same limits and not adhering to the same standards of conduct can have an unfair competitive advantage.

CONCLUSION

The sea change occurring in the manufacturing sector regarding environmental concerns is reminiscent of the change that occurred with the quality movement. Quality was first viewed as a cost and then became a necessity. Today, it is a critical element of competitiveness. Environmental concerns appear to be following the same trajectory, and companies are learning to better integrate these concerns into their core operations. Still, barriers remain, and many questions are unanswered. As with the shift toward quality, the environmental initiatives may initially require a leap of faith: "Do it before measuring what it is worth." The process of identifying, measuring, and disseminating information about the incremental paybacks that accrue will follow.

Environmental stewardship in manufacturing is poised to move into an era of competitiveness. The details of this transformation remain uncertain, but there seems little doubt that the successful manufacturers of the twenty-first century will be those for whom environmental stewardship is a primary focus.

REFERENCES

Allenby, B. R. 1994. Integrating environment and technology: Design for environment. Pp. 137–148 in The Greening of Industrial Ecosystems, B. R. Allenby and D. J. Richards, eds. Washington, D.C.: National Academy Press.

Graedel, T. E., and B. R. Allenby. 1995. Industrial Ecology. Englewood Cliffs, N.J.: Prentice-Hall.

Horkeby, I. 1993. Environmentally compatible product and process development. Paper presented at the NAE Workshop on Corporate Environmental Stewardship. August 10–13, 1993, Woods Hole, Mass.

Horkeby, I. 1997. Environmental prioritization. Pp. 124–131 in The Industrial Green Game. D. J. Richards, ed. Washington, D.C.: National Academy Press.

Klamisch, R. L. 1994. Designing the modern automobile for recycling. Pp. 165–170 in The Greening of Industrial Ecosystems, B. R. Allenby and D. J. Richards, eds. Washington, D.C.: National Academy Press.

Macve, R. 1997. Accounting for environmental cost. Pp. 185–199 in The Industrial Green Game. D. J. Richards, ed. Washington, D.C.: National Academy Press.

Rejeski, D. 1997. Metrics, systems and technological choices. Pp. 48–72 in The Industrial Green Game, D. J. Richards, ed. Washington, D.C.: National Academy Press.

Richards, D. J., and R. A. Frosch. 1997. The Industrial Green Game: Overview and Perspectives. Pp. 1–34 in The Industrial Green Game, D. J. Richards, ed. Washington, D.C.: National Academy Press.

Todd, R. (Forthcoming). Environmental measures: Developing an environmental decision-support structure. In Environmental Performance Measures and Ecosystem Condition, P. C. Schulze, ed. Washington, D.C.: National Academy Press.

The Electric Utility Industry

IAN M. TORRENS AND KURT E. YEAGER

SUMMARY

In less than a decade, major corporations around the world have progressed from making occasional token public references to selected environmentally beneficial aspects of their operations to instituting increasingly comprehensive environmental audits. This auditing has been coupled with policies and actions to achieve continuous environmental improvement, as well as public reporting of the results of such efforts.

Using the industrial-ecology model, environmental performance can be further improved by creating materials and energy flows within a larger system of linked industrial units, in which waste from one is a resource for another. Complex integrated industrial ecosystems can be created in which the environmental burden of individual participants is reduced by the optimization of both energy and materials flows. At the heart of these operations is the electric utility industry.

The main products of the electric utility industry are electric energy and the services it makes possible. To produce a kilowatt-hour and get it to the consumer, the utility company must navigate a gauntlet of potential environmental and health impacts, all of which require proactive, responsible management. Electric power companies in the United States, although at different stages of the process, have all embarked on the road to corporate environmental stewardship.

Throughout the twentieth century, electricity has been a prime agent of progress, providing the foundation for increased labor productivity, capital, and primary energy resources and allowing rapid growth in prosperity, health, and quality of life. In so doing, it has become more than just an energy alternative; rather, its efficiency and precision are now essential assets to resolving the interrelated economic, environmental, and energy-security issues facing the world today. However, the generation and delivery of electricity (and even its use) are under fire as contributors to environmental problems, both in advanced and developing countries.

Major contributions to improved environmental performance can be made by switching to cleaner primary energy sources, reducing emissions, improving the efficiency of generation, delivery, and use of electricity, and preventing pollution through better management of by-products. From the industrial ecology perspective, optimizing industrial ecosystems can involve the use of waste heat from electricity generation for residential and commercial heating and, conversely, the use of municipal waste or refuse-derived fuels as sources of energy for electricity generation. There is also considerable promise for alleviating the negative environmental impacts of other industry sectors, such as transportation, through the substitution of fossil fuels by electricity.

This paper addresses the electric power sector and industrial ecology through specific examples of the environmentally beneficial use of electricity-using technologies, or electrotechnologies, in industry, municipal waste and water management, and transportation. It also discusses how the utility industry increasingly is preventing pollution through source reduction, reuse and recycling of its by-products, and use of management tools such as waste accounting and life-cycle analysis. In the pollution-prevention activities of electric utilities, there is a significant amount of symbiosis with other industry sectors.

The trailblazing technologies of the late twentieth and early twenty-first centuries are likely to open the door to a new cycle of growth in the use of electricity. This places on the providers of electric power a particular responsibility for the environment. It also puts them at the focal point of any industrial-ecology system. Through the benefits of technological innovation made possible by electricity, electric power companies have an opportunity to be key players in worldwide economic development. Only through their continued careful attention to clean and efficient generation and delivery of electricity, however, can they play their commensurate role in contributing to improved environmental quality and sustainable development.

BEYOND COMPLIANCE TO STEWARDSHIP

Less than a decade ago, satisfactory environmental performance in the electric utility industry, as in many other industrial sectors, entailed simply complying with applicable laws and regulations. To the chief executive officer (CEO), these requirements represented an additional cost of doing business—a cost that was unfortunately growing as Congress and the Environmental Protection Agency (EPA) continued to invent new constraints in the name of environmental protection. Several events and trends (Box 1) have combined over the past 10 years to change industry's attitudes.

> **BOX 1   Factors Contributing to Industry's Changed Attitudes Regarding the Environment**
>
> - The Brundtland Commission report (World Commission on Environment and Development, 1987) forced into political prominence the concept of sustainable development.
> - Public concern over potential global climate change due principally to carbon dioxide emissions from fossil fuel use has provided support for international efforts at climate stabilization.
> - The rapid pace of economic development in Asia and the collapse of the Soviet Union have raised the profile of public concern about the actual and potential environmental degradation due to inadequate attention to environmental protection.
> - The not-in-my-backyard (NIMBY) attitude of local residents is forcing industry to reconsider its resource use and waste-generation practices (especially where hazardous materials are concerned).
> - The 1992 Rio Conference caused industrial companies worldwide to recognize these trends and to take an entirely different perspective on their environmental roles and responsibilities to national and local governments.

Today, corporations are instituting increasingly comprehensive audits of the environmental impacts of their operations and efforts to minimize these impacts. This auditing has been coupled with policies and actions to achieve continuous environmental improvement, as well as public reporting of the results of the efforts. The trend is still in its infancy but is being driven rapidly toward general adoption. Companies that seek competitive advantage through public perception of superior environmental performance are allying themselves with environmental, financial, and management consulting groups that have the expertise necessary to transform environmental performance reporting from an esoteric analysis into a mainstream corporate activity. It is a fast-moving field. The efforts of individual companies are often driven by the vision of a CEO who is convinced that his or her business will benefit from being viewed as searching out more environmentally benign ways to bring products to market and ensuring that their use, reuse, recycling, and disposal are equally environmentally sensitive.

In each industry sector, as a nucleus of "green" companies takes shape, the industry trade associations become involved in improving environmental performance, mainly by transferring to member firms methodologies for environmental auditing and reporting. Accountancy also becomes involved. The investment community is frequently involved as well, principally because of its concern about the potential future liability of companies for past, present, or future environmental damage. It is perhaps not an exaggeration to suggest that financial analysts worldwide will, before the end of this decade, study company environmental goals, practices, and performance reports with as much diligence as they today examine these firm's financial statements. Finally, governments are increasingly mandating disclosure of environmentally relevant information by companies.

The electric power sector is no exception to these general trends. Its main products are electric energy and the services it provides. To produce a kilowatt-hour and get it to the consumer, however, the company must navigate a gauntlet of potential environmental and health impacts, all of which require proactive, responsible management. Electric power companies in the United States, although at different stages of the process, have all embarked on the road to corporate environmental stewardship.

PURSUING ENVIRONMENTAL EXCELLENCE

What causes a company to try to improve its environmental performance and to communicate the results of this effort to its stakeholders? Where is the profit in developing symbiotic relationships with others in an industrial ecosystem?

In a 1991 report, the United Nations Environmental Program's Industry and Environment Office (UNEP/IEO) identified the following factors that might motivate companies to seek a greener image:

- increasing legal requirements and regulations;
- widening environmental responsibility for products and processes;
- public opinion and pressure;
- accidents or environmental events;
- the presence of a champion or visionary at the executive level;
- developments in technology and skills;
- general efficiency and quality improvement;
- competition and peer pressure;
- desire to foster a positive public image and consumer acceptance of company and products;
- desire to attract good employees;
- recognition of new markets and opportunities; and
- avoidance of cost and liability.

The UNEP/IEO report pointed to the first three factors as the most influential motivators. The first requires no elaboration. As to the second, industry is increasingly being persuaded to take responsibility for the potential environmental impacts of products during their cradle-to-grave life cycle. Companies can expect that, in the future, this responsibility will extend beyond emissions, waste, and energy-efficiency issues to encompass global efforts on environmentally sustainable development. The third motivating factor, public opinion and pressure, plays a strong role in influencing the public image of the company and in influencing governments to pass new environmental laws and regulations. The remaining factors also provide sound reasons for firms to move swiftly to embrace positive and proactive environmental stewardship.

Hedstrom and McLean (1993, p. 19) identify six "imperatives for excellence in environmental management." They argue that these imperatives, listed below, will influence competitiveness and, for some, determine survival.

- *Define policy to push the vision.* The policy should include minimal performance standards, a detailed implementation plan to translate policy into goals and target dates, and responsibilities, accountabilities, and incentives for reaching goals.
- *Measure to manage.* Measurement will not be perfect, but without useful measures, no improvement can be credibly cited. Progress in this area today is extremely rapid.
- *Communicate to establish dialogue.* This means asking stakeholders what they want to know, not merely telling them what you think they should know. It requires developing appropriate means of communicating this information. Communications should be tailored to the differing needs of stakeholders, and the communications process should be refined continuously.
- *Question "business as usual."* A leadership position in environmental performance as currently defined might not hold its standing even 5 years from now. The rapid evolution of this aspect of business requires continual improvement of environmental performance.
- *Satisfy all stakeholders.* Like any business strategy, environmental strategy can succeed only if it meets the needs of the company's stakeholders. Relationships might have to be rethought of as partnerships to meet mutual goals.
- *Integrate.* Environmental management has to be a core concern of critical decisionmakers. Environmental considerations must be built into line-management responsibility alongside considerations of efficiency, productivity, quality, and profitability. In considering the company's role in a larger industrial ecosystem, integration goes well beyond the single company to embrace those with which it does business as well as its interactions with the natural ecosystem.

The Edison Electric Institute (EEI) has compiled a menu of initiatives that it believes contribute to overall environmental excellence. In broad terms, the initiatives track Hedstrom and McLean's imperatives. EEI divides them into five groups:

- Corporate environmental commitment (includes public statement of commitment and goals; CEO leadership; environmental goals integrated

into business planning; adequate resources; and involvement of key stakeholders)
- Environmental performance measurement and reporting (includes active and thorough compliance management; prompt response to problems; development of key performance indicators; environmental auditing; and internal and external communication).
- Pollution prevention and waste minimization (includes environmentally sound procurement, use, and disposal; energy efficiency and integrated resource planning; release prevention and emergency response planning; multimedia approach to pollution control; life-cycle assessment and practices; and environmental risk management).
- Employee training and responsibility (includes awareness and technical training; clear employee goals, responsibility, and accountability; encouraging employee input on environmental improvement; and an atmosphere that encourages positive behavior).
- Environmental stewardship (includes proactive stance toward environmental issues; investment in environmental enhancements; focus on continuous improvement; and research and development (R&D) on environmental improvement).

Electric power companies are engaged in a range of initiatives along these lines. Many if not most have publicly announced environmental commitments or goals and are in the process of establishing internal environmental databases and performance measures. A number have issued formal environmental reports or have included environmental sections in their annual reports. The process is still evolving, however, in part because top management is currently much more focused on the impact that deregulation and competition have on the balance sheet.

As utility companies face up to the changing business environment and reorganize to survive and profit from it, they should incorporate the changes necessary to make environmental excellence an integral part of their business. In the words of Wilson and Greeno (1993, p. 5), "They will need to shift their environmental management from a largely functional approach to a more business-oriented perspective." Wilson and Greeno listed the following characteristics as necessary for a state-of-the-art environmental management program.

- The needs of environmental stakeholders will be addressed and satisfied through critical business processes that incorporate environmental concerns across the full product life cycle. Superior business performance processes will be defined to include environmental performance.
- The environmental vision will be part of the broader business strategic vision. Strategic planning will explicitly integrate environmental needs.
- Line managers will carry full accountability for environmental perfor-

mance, with incentive and reward systems that reflect this. They will look to much leaner corporate environmental staffs only for specialized services.
- The chief corporate environmental officer will rotate from and to line management and will have direct access to the top management levels.
- Reliable tracking and measurement will support continuous performance improvement and an aggressive internal and external communication program aimed at key stakeholders.
- As competitive pressures intensify, outreach to a wider range of partners within the industry, among the public, and in government will contribute to a reputation for environmental excellence and a greater voice in the public arena.
- Strong environmental performance will provide leverage for strategic advantage, increasing the company's ability to influence events and control its destiny.

The concepts delineated above for individual companies and industry associations are taken a step farther in industrial ecology. Industrial ecology views the industrial system as a complex web of interrelationships in which waste is eliminated by optimizing the flows of energy and materials among system participants, including the natural ecosystems that are critical participants in these flows. According to Tibbs (1992, p. 4), this will require

> designing industrial infrastructures as if they were a series of interlocking man-made ecosystems interfacing with the natural global ecosystem. Industrial ecology takes the pattern of the natural environment as a model for solving environmental problems, creating a new paradigm for the industrial system in the process.

The ideal industrial ecosystem is one that is as close as possible to a closed-loop system, with near-complete recycling of materials and cascading of energy.

THE IMPORTANCE OF COMMUNICATION

Some of the organizational implications of improving a company's environmental performance are discussed above. As with any major change in the way a company does business, the quest for environmental excellence must be led in clear and certain terms. The solid commitment of top management is essential. INSEAD, the European Institute of Business Administration, found that the commitment of top management "may be the single most important criterion for the successful implementation of good environmental practices within the company" (United Nations Environmental Program, Industry and Environment Office, 1991). Top management leadership determines whether the company will be a leader or a follower in environmental protection.

Effective communication of this environmental vision to middle managers and employees is the next most important criterion for successful cultural imprinting. A company's environmental policy, enthusiastically and openly backed by the CEO, serves to communicate the company's commitment to deal with environmental concerns. New environmental policies and goals have to be "sold" and appropriate accountability designated to all levels. If this does not happen, line employees and middle managers will see demands to make processes more environmentally sound as just another added cost of dubious necessity that makes it more difficult for them to reach their bottom-line goals. Therefore, environmental performance needs to be communicated as a key component of all other corporate goals. Monitoring and reporting of performance need to be institutionalized, and employee performance needs to be judged on how well employees integrate improved environmental performance into their other responsibilities.

Employee training on environmental matters is essential if performance improvements are to be realized efficiently and rapidly. Employee environmental training often combines broad-based awareness training and job-specific technical training. Awareness training exposes employees to the link between the company's environmental policy and implementing practices and provides an effective forum for employees to exchange ideas and suggestions regarding environmental management, responsibility, and goals. Technical training targets employees who have specific responsibilities for complying with environmental requirements or specific opportunities for contributing to the continual improvement of the company's environmental performance. Effective training is aimed at cultural change and strives to better define employee goals, responsibilities, roles, and personal accountability for achieving the company's environmental objectives.

Finally, the two-way-street aspect of training is essential to making the cultural shift. Employee input to the process of environmental compliance or stewardship should be strongly encouraged. Employee focus groups can even be helpful to management in exploring the issue of industrial ecology and its relevance to the company in the short and long term. Employees who come up with ideas that are implemented and result in improved performance should be rewarded. This type of dialogue is very effective in disseminating the culture of corporate environmental excellence.

THE IMPORTANCE OF MEASURING ENVIRONMENTAL PERFORMANCE

Environmental performance indicators (EPIs) attempt to measure a company's success in environmental management and protection, enabling it to set environmental performance targets and chart progress toward those goals. If these indicators can be made comparable across an industry, among industrial sectors, and ultimately among countries, they can be used as benchmarks. Current best practice involves the use of indicators that measure significant environmental im-

> **BOX 2   Core Environmental Performance Indicators**
>
> - Materials used
> - Primary energy used
> - Carbon dioxide emitted due to energy use
> - Emissions of different air pollutants
> - Water consumed
> - Major waste streams
> - Water effluents (biochemical and chemical oxygen demand)
> - Percentage of waste recycled
> - Percentage of recycled materials used in processes
> - Number of environmental accidents and human health impacts
> - Product impacts during use
> - Level of expenditure on environmental protection
> - Evaluation of total resource use
> - Environment-related liabilities

pact, can be self-assessed and externally verified, and are comparable either over time or with the best representative environmental standards (Deloitte Touche Tomatsu International, 1993). There is a fairly large set of frequently used core EPIs (Box 2) that can be used to assess environmental improvement and to rate a firm's performance within the industry sector.

Two models for corporate environmental performance measurement and reporting have emerged (Elkington and Robins, 1994). The Anglo-Saxon model, favored by North American and United Kingdom companies, is based mainly on an inventory of emissions and management practices. The Rhine model, used by German and Scandinavian companies, takes the practice a step farther to reporting on a company's eco-balance, which is based on a life-cycle accounting of environmental impacts (positive and negative) associated with the raw materials used as well as the company's final products.

THE UNIQUE CONTRIBUTIONS AND CAPABILITIES OF ELECTRICITY

Throughout the twentieth century, electricity has been a prime agent of progress, providing the foundation for increased productivity of labor, capital, and primary energy resources, and for rapid growth in prosperity, health, and quality of life. In so doing, it has become more than just an energy alternative; rather, its efficiency and precision are now essential assets to resolving the interrelated economic, environmental, and energy-security issues facing the world (Yeager, 1994).

However, the generation and delivery of electricity (and even its use) are under fire as contributors to a range of environmental problems in both advanced and developing countries. These include the conventional impacts of air, water, and soil pollution; the issues of safety and waste disposal related to nuclear power; possible health impacts of exposure to electric or magnetic fields; and emissions of greenhouse gases, which could affect global climate patterns. The challenge, therefore, is to demonstrate the unique capabilities of electricity versus other forms of energy in terms of versatility, conversion efficiency, sustainability, and control over potential environmental impacts.

Several options are currently available to the electric utility industry to improve its environmental performance. Major contributions to this effort can be made by switching to primary energy sources with fewer environmental impacts; improving the efficiency of the generation, delivery, and use of electricity; preventing pollution through better management of by-products; and substituting the direct use of fossil fuels by electricity-using technologies, or electrotechnologies, at the point of end use. Figure 1 summarizes the opportunities available to improve environmental quality in the electric utility industry.

Much has been written on the question of fuels and efficiency. There is considerable promise both for generating and using electricity more efficiently and for deriving environmental benefits for other industry, transportation, and residential and commercial consumers through substitution of electricity for the direct use of fossil fuels (referred to as beneficial electrification) (Yeager, 1994). From the industrial-ecology perspective, examples of cross-sectoral optimization include the use of waste heat from electricity generation for other purposes, such as industry or residential and commercial heating and, conversely, the use of municipal waste or refuse-derived fuels as a primary energy source for electricity generation.

A frequently cited example in industrial ecology is the cooperation in Kalundborg, Denmark, between the Danish utility Asnaes and a web of other industries, including an oil refinery, a biotechnology plant, a gypsum wallboard manufacturer, a sulfuric acid producer, cement producers, a district heating system, and local agriculture, horticulture, and pisciculture (Figure 2) (Grann, 1997). Water, energy, chemicals, and organic materials flow from one company to another, decreasing waste production as well as air, water, and land pollution.

It is undoubtedly well-known how electricity can be produced from its primary energy sources and used cleanly and efficiently. This paper therefore bypasses those topics and concentrates in greater depth on two aspects of electricity use with broad implications of industrial ecology. The first is the beneficial role electricity can play in solving environmental problems through substitution of electrotechnologies in the industrial, transportation, commercial, and residential sectors. The second involves the pollution prevention aspect of electric utility business operations.

Beneficial Electrification

Environmental protection will continue to be a priority for electric utility companies. However, in addition to controlling pollution and aiding end-of-

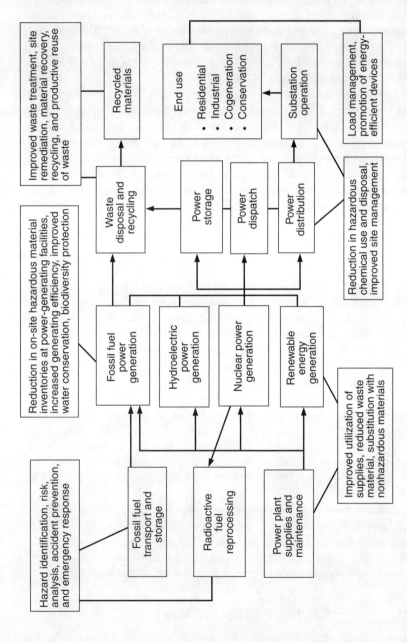

FIGURE 1 Opportunities to improve environmental quality in the electric utility industry. SOURCE: Yeager, 1994.

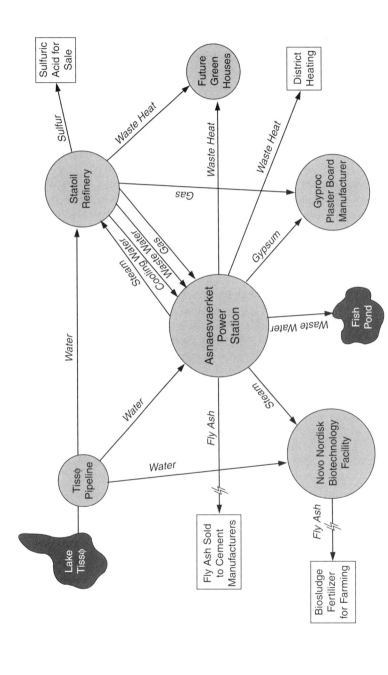

FIGURE 2 Schematic of industrial ecosystem involving electric utility at Kalundborg, Denmark. SOURCES: Grann, 1997; Tibbs, 1992.

pipe cleanup strategies, U.S. utilities are increasingly emphasizing the proactive use of electricity as a way to address the environmental concerns of their customers.

This approach turns an environmental debit into a business credit. From the perspective of environmental quality, electricity might be part of the problem, but it is also consistently a part of the solution. Much progress is being made in electrotechnologies, from home appliances to industrial processes to computers. Better use of electricity has already saved the United States at least $21 billion by reducing the operating costs of new power plants (EPRI Journal, 1992).

Also, it is not simply a question of using less electricity to produce the same quantity of goods and services. There are many instances in which industrial use of electricity might be more efficient and less polluting than direct use of a fossil fuel in the manufacturing process. Such beneficial electrification is likely to improve environmental quality not only by reducing aggregate pollutant emissions and impacts associated with economic activities, but also by reducing carbon dioxide emissions, a greenhouse gas linked to climate warming.

New, environmentally cleaner electrotechnologies will mean new business opportunities. For example, plasmas can be used to destroy medical waste; electron-beam treatment can be used to eliminate toxic gases from municipal and rural wastewater; and electrolytic ozone treatment of effluents can reduce color and toxins during waste treatment. Transportation is another area in which electrification can improve capacity utilization. The following sections describe ways that beneficial electrification can lead to win-win-win situations for energy production and use, the environment, and sustainable economic development.[1]

Industrial Electrotechnologies

Opportunities for increased use of new electrotechnologies abound in virtually all segments of industry and are likely to be developed substantially over the next 20 years. Even though many industries are relatively insensitive to the costs of energy, there can be substantial incentives to replace traditional processes with new electrotechnologies that reduce primary energy consumption, decrease emissions to the environment, and increase productivity.

An example of the multiple benefits available through electrification can be seen in the use of freeze concentration in the dairy industry. The dairy industry is the largest food-industry user of energy for concentrating raw products. Much of the equipment currently in use—typically involving the use of thermal evaporators—is antiquated, inefficient, and consumes relatively large quantities of fossil fuels at the point of use. Freeze concentration, a recently developed technology, uses electricity-based vapor compression to freeze out water. Freeze concentration is more energy efficient than the conventional natural-gas-based process and yields improved taste and aroma, reduces spoiling, and lowers cleaning costs at the processing site.

Beneficial electrification is also evident in a new generation of advanced motors and adjustable-speed drives. Motors represent about 70 percent of industrial electricity use, and many can operate only at a narrow range of speeds, creating energy losses and process inefficiencies. Advanced motor designs will permit better integration of the motor and power supply combination. Electronic adjustable-speed drives allow motor speeds to be precisely varied without loss of efficiency or damage to the motor, saving energy and improving process control.

The energy, environmental, product-quality, and productivity benefits of evolving electrotechnologies have been demonstrated in the following industrial applications:

- *Infrared heating.* The use of infrared heating for industrial drying and curing is an alternative to conventional gas ovens for setting finishes on many products, including painted car bodies and home appliances, printed paper, coated steel and aluminum coils, and painted or varnished hardboard and particleboard. Infrared processing consumes only about half the total energy of conventional gas-fired convection heating, eliminates gas-combustion emissions, and lowers emissions of carbon monoxide, carbon dioxide, and volatile organic compounds (VOCs).
- *Electric-arc furnace.* Electric-arc furnaces melt steel by passing electric current directly through the raw metal. They use less than half the total energy resources of the traditional coke-fired steel-making process, and there are many fewer emissions. Arc furnaces do not release VOCs, such as benzene, which are emitted by coke ovens. Electric-arc furnaces are more flexible in size and more economical to build and operate, and they are an essential tool for metal recycling, because they can melt 100 percent of scrap steel.
- *Foundry casting-sand reclamation.* High-intensity electric infrared emitters can clean the spent sand used by foundries to make molds and cores for metal castings. The process reduces the volume of foundry sand being dumped into landfills, currently about 7 million tons annually (Foundry Management and Technology, 1991). The on-site electric heat source produces no emissions, and the recycling process saves oil and reduces carbon dioxide emissions by eliminating the need to transport more sand.
- *Ultrasound textile dyeing.* Studies show that an ultrasound dyeing process can cut dye time in half and increase color and yield in a wide range of fibers. Environmental benefits include reduced energy consumption, less dye waste, and a reduction of auxiliary chemicals and scouring agents.
- *Process-water recovery.* The food industry, the largest economic sector in the United States, faces critical water usage and water management issues in processing plants. Use of electrically driven membrane-separation technologies offers an attractive solution to treating industrial-process water streams under tougher environmental regulations.

Electrotechnologies for Municipal Water and Waste Treatment

Environmental concerns, rising costs, and new regulations are challenging water and wastewater utilities and health care providers across the country. Recognizing the importance of addressing these challenges, the Electric Power Research Institute (EPRI) is working with the American Water Works Association Research Foundation; the Water Environment Research Foundation; several dozen electric, water, and wastewater utilities; and government and research institutions to identify innovative solutions based on new electrotechnologies. This collaboration is providing more effective treatment processes and is reducing water treatment by-products, environmental impacts, and energy and operating costs.

One of the key steps in treating drinking water or wastewater is disinfection. Chlorination is the most common disinfection method now in use, but over the last 2 decades its efficacy has been questioned. Starting in June 1993, water utilities faced new requirements for the reduction of viruses and cyst-forming parasites. Since 1996, they have also had to comply with new regulations for limiting by-products of disinfection in potable water. Several new electrotechnologies offer attractive alternatives to traditional water-treatment methods:

- *Ozonation.* For disinfection of parasites such as Giardia, ozone can be 100 to 300 times more effective than chlorine, while producing fewer disinfection by-products. Ozonation also destroys many of the organics in drinking water that can produce objectionable taste or odor. In an EPRI-funded project with Union Electric Company and the St. Louis Water Company, ozone has also been shown to reduce herbicide residues in water supplies.

    Ozone is produced from oxygen when an electric charge passes through air or when oxygen is passed between concentric tubular electrodes. The ozone-enriched gas is then bubbled through water, and the residual ozone is destroyed or recycled. Although the results are highly effective, using ozone to disinfect drinking water is about twice as expensive as using chlorine. Current research is directed at improving the efficiency and reducing the cost of ozonation.

    The higher cost of ozonation can be offset by reducing the amount of energy used in other water-treatment processes. Potable water-treatment facilities are particularly attractive candidates for introducing more energy-efficient technology. Installation of adjustable-speed drives, better instrumentation and control systems, and other measures can reduce energy use substantially, usually with short payback periods.
- *Ultraviolet treatment.* For disinfection of wastewater, ultraviolet (UV) treatment offers several technical advantages over conventional methods as well as competitive costs. A UV disinfection system consists of a network of fluorescent lamps with special quartz glass that transmits UV light. This method of treatment eliminates the need to store or handle

dangerous chemicals and greatly reduces chlorination by-products in treated wastewater discharges. Although exposure to the lamps quickly kills bacteria and viruses, it does not readily inactivate parasites and thus is not a good candidate for treating surface water. However, it might be an option for disinfecting groundwater, where parasites are not a problem.
- *Electron-beam disinfection.* Electron-beam disinfection is an experimental technology that disinfects wastewater and destroys organic compounds. The process involves bombarding water with high-energy electrons from a particle accelerator. The electrons and the chemical radicals created in this process destroy microorganisms and organic molecules in wastewater. Tests of the technology suggest that the process holds promise, but more research is needed.
- *Sludge treatment.* Sludge produced as a by-product of water or wastewater treatment is difficult to manage, mainly because of its high water content. About 50 million tons of sewage sludge are produced nationwide each year, with a typical moisture content of about 75 percent (Metcalf and Eddy, 1991). Better drying techniques can lead to lower disposal costs and increased use of sludge as a soil amendment or fertilizer. Several promising electrotechnologies for drying sludge are under development, including a mechanical freeze-thaw process.
- *Desalination and water reuse.* In coastal areas with limited freshwater resources, electrotechnologies can play a direct role in providing potable water, either through desalination of ocean water or through reclamation of wastewater. Both processes are energy intensive and can make use of waste heat or electricity generated at power plants. An additional factor making it advantageous to locate power and desalination plants together is the opportunity to share intake and discharge structures. Two technological developments offer opportunities for combined facilities: Low-temperature, multieffect distillation can make efficient use of relatively low-temperature steam from a power plant for desalination; and advanced reverse osmosis systems are bringing down costs to the point where they promise to be commercially competitive.

Medical-Waste Treatment

Environmental and health concerns have heightened the public's anxiety about medical waste and its disposal. EPA estimates that U.S. hospitals generate more than 2.5 million tons of solid waste annually, about 15 percent of which is infectious (U.S. Environmental Protection Agency, 1994). Municipal landfills are prohibited from accepting infectious waste, so hospitals treat it on site (often by incineration) or ship it to hazardous-waste disposal facilities.

New electrotechnologies can either destroy infectious waste without incineration or disinfect and shred the waste, permitting disposal in municipal landfills.

EPRI is conducting collaborative projects to demonstrate the feasibility of a number of these electrotechnologies:

- *Electropyrolysis.* Electropyrolysis involves heating waste electrically to temperatures greater than 500°C in the absence of air to convert the waste into nonhazardous ash and flammable gas. The gas can be drawn off and burned at high temperature in a controlled atmosphere.
- *Microwave treatment.* Microwave disinfection is accomplished in an enclosed, trailerlike container in which the waste is shredded and then disinfected by exposure to microwaves and sustained high temperatures. The treated material, similar to confetti, is shipped to a municipal landfill. Energy consumption is modest, and other benefits include a 90-percent reduction in volume and lower disposal costs.
- *Plasma processing.* In plasma processing, electricity is used to heat a gas mixture to a partially ionized state known as a plasma. The plasma is then used to heat the waste to about 1600°C in the absence of air. The waste is destroyed and the slag and gases that are formed can be flared or burned to provide steam.

Transportation

The move toward electric vehicles (EVs), which are typically 40 to 60 percent more energy efficient than gasoline vehicles, is gaining momentum worldwide. Progress is constrained by the practicality of battery technology, but it might be facilitated by the market demand for hybrid electric and combustion-powered vehicles needed to meet urban air-quality standards.

The benefits of electric transportation are clear. Road vehicles and transit systems powered by electricity offer clean, quiet, and reliable alternatives to those powered by the internal combustion engine. These attributes are especially important for urban areas affected by air quality problems. Despite strict smog controls, nearly 100 cities in the United States still fail to meet federal clean-air standards, and vehicle emissions are a leading cause.

In response to state laws that require automakers to produce increasing numbers of zero-emission electric vehicles, EV sales are expected to grow dramatically in the new millennium. A bellwether California law requires that by 1998, 2 percent of new cars produced for sale there be zero-emission vehicles. This figure will rise to 10 percent in 2003. New York and Massachusetts have approved similar rules. The Electric Transportation Coalition estimates the rules will contribute to the sale of at least 49,000 electric cars in California and New York in 1998, 122,500 cars in those two states in 2001, and 245,000 cars in 2003.

Electric vehicles are the only option available to meet zero-emission requirements. They produce no tailpipe emissions and are as much as 10 times cleaner than the most advanced gasoline-powered vehicle, even when power plant emissions associated with the production of electricity for EV use are taken into account.

Wider use of EVs and electrified public transportation is expected to reduce substantially levels of important air pollutants, particularly nitrogen oxides, carbon monoxide, and particulate matter. EVs waste no energy and produce no pollution while idling in traffic. During braking, their motors automatically become generators, recovering energy that is used to recharge their batteries and further efficiency. Additional improvements in efficiency are expected.

Electric vehicle battery recharging will occur largely at night, when it is most convenient for EV owners and most efficient for utility generating systems. There is enough idle capacity available overnight to meet the recharging needs of millions of EVs.

Here and elsewhere in the world, the EV holds great promise for reducing emissions that are linked to concerns about global warming, depending on the fuels used for electricity generation. Carbon dioxide is the principal compound linked to global climate concerns, and exhaust emissions from conventional vehicles are a significant source of this greenhouse gas. Although EVs themselves produce virtually no emissions, the power plants that generate their electricity do. Even so, widespread use of EVs could reduce greenhouse gas emissions by 30 percent or more, and perhaps by substantially greater amounts as more advanced technologies come on line.

A partnership among the federal government, the nation's "big three" automakers, and EPRI has created the U.S. Advanced Battery Consortium, a 4-year, $260 million research program whose goal is to improve EV performance through improved batteries. The project's target for the year 2000 is to develop commercially available EV batteries that can power a car for 200 to 300 miles without recharging.

To ensure that the infrastructure is in place to support wider use of EVs, the electric power industry is supporting development of connecting-station technologies and standards and equipment for charging EVs, such as quick-charge stations that can extend an EV's range by as much as 60 miles with a 6- to 12-minute charge.

The use of electricity for transportation is not, however, limited to EVs. Advanced rail transportation is a reality in Europe and Japan and could be economically viable in several highly populated corridors of the United States. Electrically powered vacuum MAGLEV technology, for example, promises to be able to link cities hundreds of miles apart with transit times measured in minutes.

HOW ELECTRIC UTILITIES ARE ADDRESSING POLLUTION PREVENTION

The electric utility industry produces a variety of by-products in the process of generating and distributing electricity and servicing its customers. These by-products span a wide range in terms of their qualities and rate of generation, chemical and physical form (solids, liquids, and gases), and management methods. In the first category are high-volume by-products consisting mainly of coal ash and solid and liquid by-products from control of gaseous emissions that are produced in multiple tons per year. The second category consists of noncombustible wastes produced in low volumes. Examples include liquid-filled fuses, asbestos, solvents, paint smudges, boiler-cleaning waste, various blowdown streams, and ash-pond discharges.

The industry has historically relied on landfills and ponds as its primary solid and liquid waste-management strategy. However, utilities are turning to new strategies, including pursuing markets for their by-products through brokers, waste exchanges, and recycling or reuse; generating less material; and using more environmentally compatible materials.

Several representative case studies of pollution-prevention practices in the electric utility industry are summarized below. In addition, two other important activities under way in the industry are described. The first is the development of a life-cycle cost-management methodology to assist in making intelligent decisions regarding chemical and materials purchases. The second is the development of a waste-accounting methodology to track the progress of pollution-prevention efforts.

Use of High-Volume By-Products

High-volume by-products are generated as a result of the air emissions control system at a power plant—electrostatic precipitators, baghouses, or flue gas desulfurization (FGD) systems. Currently, the industry produces about 90 million tons per year of coal ash, bottom ash, boiler slag, and FGD by-product—enough material to fill a football stadium to a height of 9 miles (American Coal Ash Association, 1993). Given projections that coal use in the United States might increase substantially in the next 10 to 15 years and the growth anticipated for FGD systems because of new Clean Air Act legislation, the challenge to manage this increasingly large quantity of by-products will grow substantially.

Currently, about 20 percent of electric utility industry high-volume by-products are used in commerce. The most commonly used by-product is fly ash, which can substitute for cement in concrete. Utilization trends have remained relatively constant over the last several years. Although there are many uses for coal ash, given its pozzolanic and, in some cases, its cementitious properties, most potential markets are relatively small. The following examples focus on some new, potentially large markets for coal ash.

- *Highway construction.* Highway construction represents a major potential market for coal-combustion by-products. This usage to date has been limited, however, primarily because state highway departments and contractors are unfamiliar with coal ash in these applications. In addition, state environmental agencies that grant permits need information on the potential groundwater impacts of the ash constituents. Finally, the materials that coal ash would replace—cement, sand, soil, and aggregate—are typically locally available at relatively low cost, and the suppliers are structured as vertically integrated industries. Thus, coal-combustion by-products have a difficult time cracking the market unless there is a strong cost advantage.

EPRI and several of its member utilities have sponsored five highway demonstration projects using coal ash in several different applications, including as a road subbase, a base course, in embankments, and as a high-cement replacement in concrete-base course. The projects were structured to show the technical acceptability of ash in these applications, as well as to examine any short-term effects on groundwater. One utility in Pennsylvania used coal ash and stabilized FGD by-product rather than conventional fill materials in an embankment for an interstate highway. It documented savings of over $600,000 for a 1,500-foot highway section using 353,000 tons of coal ash. These demonstration projects have served as test cases in developing highway design and construction manuals on using coal ash (Electric Power Research Institute, 1988; Patelunas, 1988).

- *Autoclaved cellular concrete.* Autoclaved cellular concrete (ACC) is a lightweight concrete with no coarse aggregate. It is produced by mixing Portland cement, lime, aluminum powder, and water with a large proportion of a silica-rich material. In many countries, sand is the silica source. EPRI and the electric utility industry are investigating substitution of coal fly ash for the sand (as is done in the United Kingdom). Typically, fly ash can account for up to 75 percent of the solid material that makes up ACC. Initially, the focus of this effort has been on the production of standard masonry blocks. Using this material produces blocks that weigh one-quarter that of conventional masonry blocks, have thermal insulation properties (allowing insulation costs to be avoided in some climates), have self-leveling properties, and are fire and rot resistant. An additional advantage is that conventional carpentry tools can be used during construction. The technology has been commercialized in over 40 countries, with about 160 plants in operation.

    The challenge is to introduce this technology in the United States and to displace existing concrete masonry blocks, and possibly even some wood, in building construction. EPRI and the electric utility industry are now sponsoring a series of demonstrations of the technology at eight power plants using a mobile pilot plant (Sauber, 1992). At each site, the local construction community can witness the production process firsthand and see how the blocks are used in field demonstrations. The ultimate goal is to stimulate sufficient interest and markets that the business community will invest capital to operate commercial-scale plants in the United States using recycled coal ash.

- *Aggregate production.* Although the largest use of coal ash is as a substitute for cement in concrete, concrete contains only about 20 percent cement. Most of the remainder is aggregate and water. Therefore, manufacture of aggregate from coal ash would represent a much larger market in the concrete industry. An example of a recent entry into this market is the Aardalite process. It consists of adding together coal ash, water, lime, and additives and passing the resulting mixture through a rotating disc pelletizer, which forms pebblelike material that can be sized according to need. A 24-hour steam cure is the final step. A full-scale plant is currently in operation in Florida and uses about 150,000 tons of coal ash per year.
- *Other uses.* Some promising new applications are currently in the research stage, including using these by-products as filler materials for plastics and metals. Not only could these fillers be produced more cheaply than the substances they replace, but the ash would add desirable properties to the final product. In aluminum, the ash improves the metal's machining and sound-damping characteristics, making it a good candidate for engine casings. In plastics, the ash could reduce the need for expensive matrix components such as resins, or reduce the amount of elastomers in rubber.

The electric utility industry and EPRI are aggressively pursuing existing markets for coal ash and investigating new opportunities through research. Although no national goal for coal ash use has been established by the industry, EPRI, the American Coal Ash Association, and the Edison Electric Institute have been active in promoting coal ash as a useful by-product and in attempting to remove institutional barriers to its use (Brendel and Kyper, 1992).

Management of Low-Volume or Noncombustion Waste

The electric utility industry is also actively pursuing waste-minimization and recycle-reuse options for a variety of other waste materials generated as part of electricity production and distribution (Electric Power Research Institute, 1993). The major motivating factors for electric utilities in these initiatives include increased corporate responsibility for reducing waste generation, cost reduction, and responsiveness to a growing number of state pollution prevention initiatives. Many states have designated as hazardous a wide variety of waste materials; therefore, costs of landfill have increased substantially, and potential long-term liability remains an issue for waste that requires disposal. Among wastes being considered for recycling or reuse are:

- *Antifreeze (ethylene glycol).* Electric utilities have large vehicle fleets for servicing their customer territories. A large EPRI member utility initiated a recycling program for its antifreeze and now saves about $90,000 annually. The spent antifreeze is shipped over 400 miles for redistillation and reconstitution at a cost of $1.50 per gallon versus the replacement cost of

$8 per gallon. Further cost savings (not quantified) were realized from avoiding disposal. The utility even provides an antifreeze recycling service for its local community.

Another EPRI member utility takes the antifreeze from its vehicle fleets, filters it, adjusts the pH, and then reuses the antifreeze in an ethylene glycol air heater at one of its power plants. The capital investment was under $3,000, and operating costs are less than $1,000 per year. Over the next 10 years, the utility estimates it will save $380,000 by avoiding antifreeze disposal and replacement costs.

- *Boiler chemical-cleaning waste.* Electric utility boiler tubes must be cleaned periodically to remove iron and copper deposits that form on interior surfaces, impeding heat transfer and diminishing power plant efficiency. The types of solutions commonly used to clean boiler tubes are acidic in nature (e.g., hydrochloric acid). Utilities typically clean their boiler tubes every 1 to 5 years. The volume of undiluted cleaning waste is usually about 125 gallons per megawatt of electricity produced per tube. When this material is combined with several boiler volumes of rinse water, the total volume can be several hundred thousand gallons for a large power plant. Although on an annual basis this represents less than a gallon per minute, in practice the material is generated in a short time and therefore must be handled in large volumes.

Because of its pH or chromium content, some of this cleaning waste exhibits hazardous-waste characteristics as defined by the Resource Conservation and Recovery Act, considerably increasing disposal costs (Lott et al., 1989). Therefore, from a pollution-prevention point of view, options that minimize waste production, such as source reduction, substitution, or recycling, are preferred.

One innovative electric utility company has initiated a program that eliminates disposal of the cleaning waste and provides a useful product to another industry. The utility treats the metal-bearing solution with lime to precipitate the metals that are in solution, primarily copper and iron. The clean liquid is then returned to the municipal wastewater system, and the resulting sludge is sent to a copper smelter in Arizona. The smelter recovers the copper value and uses the lime as a flux in the smelting process. The utility saves an estimated $400,000 per year by avoiding treatment and disposal costs.

Other utilities have reduced their frequency of cleaning, which is a form of source reduction. Careful control of boiler-cycle water chemistry can reduce the amount of metal deposition in boiler tubes and therefore reduce the need for cleaning. Another option that appears promising is reuse of the solution in the FGD system. Laboratory tests have shown that the addition of chemical-cleaning waste slightly improves sulfur dioxide removal and increases limestone utilization while having no measurable neg-

ative impact on overall FGD system operation (Behrens and Holcome, 1992).
- *Petroleum-contaminated soil.* Petroleum-contaminated soils are the result of leaks from aboveground and underground storage tanks. Such leaks are usually managed by removing the contaminated soil and transferring it to a secure landfill. As an alternative, one utility has found that the soil can be incorporated into asphalt to pave light-duty roads near its facilities. This practice has saved $350 per cubic yard in treatment costs for the contaminated soil, or between $270,000 and $385,000 per year, assuming the paving of one road per year.
- *Spent solvent.* Utilities generate small quantities of spent solvent waste from routine operations such as parts cleaning, paint stripping, and vehicle maintenance. Because solvent waste is typically classified as hazardous, management practices are costly.

Many utilities have already initiated solvent minimization programs. Typical steps include source reduction (i.e., using less solvent), substitution of less-toxic materials, and recycling. In terms of waste reduction, using less solvent is the simplest and often the least-expensive option. Worker training and good housekeeping have been shown to be effective methods. Limiting the availability of solvents at facilities is another way to reduce usage. Electric utilities are now seeking to specify and formulate nonhazardous universal solvents to replace the wide variety of products currently in use.

A variety of nontraditional substitutes are available for halogenated organic solvents. These include sodium carbonate or sodium phosphate solutions, emulsion cleaners such as mineral spirits, and organic cleaners such as citrus-based D-limonene. Substitutes are generally less toxic but are also generally less effective, especially in electrical contact cleaning. Mechanical cleaning can, in many instances, be substituted for chemical cleaners. Extra "elbow grease" is an effective substitute, as is sand or bead blasting. Mechanical cleaning works well for paint stripping but is less satisfactory for degreasing and use on sensitive materials such as wood, plastic, or soft metal.

A survey of electric utilities conducted by EPRI found that 46 percent of respondents employ some type of solvent recycling. In most cases, recycling is less expensive than disposing of spent solvent. Recycling can be done in three ways: (1) using an on-site batch or semicontinuous distillation process; (2) contracting with a commercial solvent recycler; or (3) using contractors who provide solvent and parts-cleaning equipment.

EPRI has developed specifications for solvents in three categories: electrical equipment and parts cleaning, paint stripping, and burner tip cleaning. It has also recently completed a study with five utilities to examine

the effectiveness of approximately 15 nonhazardous solvents for a variety of applications and facilities.
- *Paint and paint-related waste.* Paint and paint-related waste represent some of the most frequently generated hazardous wastes in the electric utility industry. In addition to paints and paint smudges, these include outdated (off-specification) paints, empty paint containers, solvent- and water-based equipment-cleaning residues and paint-removal and surface-preparation residues. Simple management options are available to minimize or even eliminate paint wastes. Some utilities focus on keeping painting equipment in use to reduce the frequency of cleaning, whereas others focus on maintaining the proper inventory of paint to reduce the generation of outdated paint and contaminated containers.

  Utilities are reexamining the frequency with which equipment is painted and are seeking alternative coating materials to eliminate the need for painting. Utilities are also seeking alternative methods for paint removal and surface preparation. Some companies have successfully removed paint by blasting with dry-ice pellets or commercial blasting materials that can be separated from removed paint and then reused. Efforts continue to quantify the costs of the available alternatives for paint removal and surface preparation.
- *Biofouling waste and wood waste.* Biofouling waste includes plant species such as kelp and animal species such as Asiatic clams and shad. One utility has successfully composted shad bodies with waste wood, thereby eliminating two waste streams, in a demonstration project called Fish and Chips. Each spring, the utility collects about 300 tons of shad from its intake screen following an annual winter dieback of the fish population. It also has a steady stream of wood waste from construction projects and from routine sources such as shipping pallets and reels. Composting these wastes using native bacteria presents an attractive option, because it can deal with both waste streams. The wood is first chipped and then serves as the bulking agent to ensure there are air pockets for aerobic degradation. In addition, it is likely that the fine wood particles serve as a carbon source for the bacteria. The fish provide moisture, nitrogen, and other nutrients, in addition to serving as a carbon source for the bacteria. Windrows are established, and the two waste streams are biodegraded at temperatures sufficient for pathogen control. The utility estimates an annual savings of $58,000 for disposal of the fish, and a potential one-time savings of $480,000 for chips from a plant construction project.

The preceding examples are just a few of the new approaches being tried by utilities to improve their environmental performance. A group of 60 EPRI member utilities have estimated cost savings from implementing these options ranging from $65,000 to $140,000 per utility annually, even without allowing for reduced potential future liability (Electric Power Research Institute, 1993).

TOWARD A MORE SYSTEMATIC APPROACH: LIFE-CYCLE COST MANAGEMENT, WASTE ACCOUNTING, AND RISK MANAGEMENT

The electric utility industry is moving to a more systematic approach to managing waste materials through life-cycle cost management, waste accounting, and risk management.

- *Life-cycle cost management.* Utilities buy, use, and dispose of thousands of different chemicals and materials each year. Price has traditionally been the primary motivating force behind purchase decisions. However, the purchase price of a product or material might represent only a small fraction of its full cost. The cost of use and disposal often far outweighs the initial purchase price. Basing purchasing decisions on the initial cost of a product or material can lead to much greater costs over the long run. EPRI has defined life-cycle costs as all costs associated with a product, material, or process from purchase to potential post-disposal liability. Life-cycle cost management involves defining and characterizing costs, estimating and tracking cost elements, using these estimates in appropriate decisions, and coordinating these decisions across multiple individual and functional areas in an organization.

    Management of waste by employing life-cycle cost analyses can save money, encourage pollution prevention, reduce environmental impacts, and create new alternatives and solutions. EPRI is working with several utilities in developing methods, educational tools, cost worksheets, and computer software that will help utility staff make more informed decisions. These utilities have been assessing life-cycle costs associated with products such as batteries, utility poles, solvents, and paint and paint-removal waste. Available software packages and workbooks can structure life-cycle cost analyses, perform calculations, prompt the user with relevant advice, store data and analysis results, and allow multiple users to work together efficiently on a single analysis.

    Figure 3 shows an example of the application of life-cycle cost analysis. Two solvents are being compared: trichloroethane (TCA) and a citrus-based alternative. Using purchase cost only, TCA is the clear winner, but add use and disposal costs for 1 year of cleaning services, and the margin between the two choices becomes quite small. It falls within the uncertainties in the future price and disposal cost of TCA or in the amount of labor and materials needed to provide equivalent cleaning performance by the citrus-based product. The electric utility that performed this analysis carefully tested each cleaning method and concluded that the extra labor and materials needed for the citrus-cleaning process would exceed by no more than 5 to 20 percent the costs of comparable TCA cleaning, depending on the application. On the basis of these tests and a careful review of

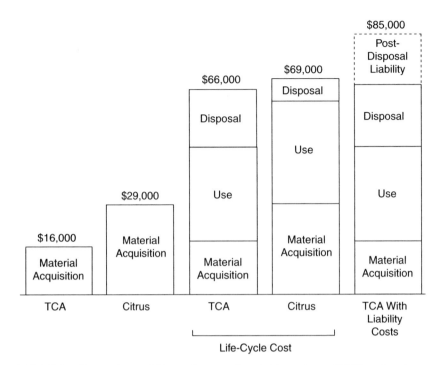

FIGURE 3 Comparison of life-cycle costs for trichloroethane (TCA) and a citrus-based solvent. SOURCE: Electric Power Research Institute, 1994.

the potential liabilities, the utility decided to make the switch to the citrus-based solvent.

- *Waste accounting.* To determine what waste a facility produces, the quantity of waste generated, and the amount of progress toward reducing waste volumes, an effective waste-accounting methodology is imperative. EPRI is working with electric utilities to develop an inexpensive, field-applicable mechanism for organizing and collecting waste-generation data at utilities. The methodology has been tested at three different types of facilities and has been coded for software implementation. This software (ASAPP, or Accounting Software Application for Pollution Prevention) is now available. It tracks the movement of materials through a company from the time they are designated as waste to their ultimate disposal. It facilitates environmental reporting, the assessment of pollution-prevention opportunities, and performance tracking. Further development of the software, currently in progress, will carry the tracking back to an earlier stage—the initial acquisition of the material by the company. This refinement will enable ASAPP to be more effective as a tool for assessing source reduction and substitution options.

- *Risk management.* For utilities to choose among waste-minimization or pollution-prevention options, they must understand the risks associated with each option. EPRI has developed a risk-management tool to aid management of noncombustible waste (NCW). Called the NCW Manager, the tool allows users to compare environmental and safety risks, direct and indirect costs, and potential liabilities for each option. EPRI is developing modules of NCW Manager to address specific utility wastes. A module on paint waste recently became available, and a module on boiler chemical-cleaning waste is being prepared.

Plans are under way to combine these pollution-prevention reports, worksheets, and software tools in a pollution-prevention workstation, which will allow coordinated use of all tools with single entry of utility data. The development of these kinds of guidelines and tools will help utilities move pollution-prevention activities from a research, assessment, and specific-application mode toward sustained, routine business practice. Eventually, these activities will become an integral part of continually improving environmental performance.

Full-Cost Accounting

Full-cost accounting is a development of life-cycle analysis and waste accounting. This approach attempts to price goods and services to reflect their environmental costs (i.e., the environmental externalities of their production, use, recycling, and disposal). Full-cost accounting is highly controversial, difficult to implement, and subject to a high degree of uncertainty and potential bias. However, because of its potential for integrating environmental considerations into economic decision making, and thus taking rational steps toward environmental improvement, it cannot be ignored.

Some companies see full-cost accounting as providing a competitive edge, like Volkswagen in Germany, which will recoup its Golf automobiles at the end of their useful life. Firms taking this approach must build the recycling or disposal cost into the initial purchase price. Meanwhile, governments and regulatory bodies are attempting to include full-cost accounting in a rudimentary form through environmental taxes and by weighing concerns about externalities when deciding to build new energy facilities. The key will be to continue building knowledge and consensus, develop new tools like life-cycle analysis, and target scientific investigation to produce more reliable information on environmental costs. Popoff and Buzzelli (1993, p. 69) observed that "when implemented correctly, full-cost accounting will improve environmental performance more than any other action, program, or regulation in place today."

In the case of the utility industry, pooling resources with regulators to try to arrive at a reasonable consensus about the best ways to internalize environmental costs could be of long-term benefit to the industry. Research on valuing environmental damages should also be consistent with developing a market-based approach to environmental regulation, that is, reducing the environmental impacts at the lowest cost to society.

CONCLUSION

Electricity can play a vital role in achieving economic growth and sustainability in a way that does not sacrifice environmental quality. The ability to convert a wide range of raw energy sources into clean, efficient power for a host of applications gives electricity unique advantages. The trailblazing technologies of the late twentieth and early twenty-first centuries are likely to open the door to a new cycle of growth in the use of electricity. Although it is difficult to see exactly how this new era will unfold, it can be said with some certainty that larger forces—even global forces—will shape electricity's role over the next 50 years. Clearly, scientific and technological developments outside the power industry—in materials, biotechnology, telecommunications, and other fields—will affect this industry. Previous electricity growth cycles, such as that which occurred just after World War II, indicate that it is the innovation during the first decade or two of the cycle that creates the growth surge.

A new cycle of growth in electricity use places on the providers of electric power a particular responsibility for environmental excellence. It also puts them at the focal point of any industrial-ecology system underpinning a twenty-first-century economy. Through the benefits of technological innovation made possible by electricity, electric utilities have an opportunity to be key players in economic development worldwide. Only through their continued careful attention to the clean and efficient generation and delivery of electricity, however, can they play their commensurate role in contributing to improved environmental quality and sustainable development.

NOTE

1. More examples can be found in the April/May 1992 issue of *EPRI Journal*.

REFERENCES

American Coal Ash Association (ACAA). 1993. Coal Combustion By-Product Production and Use, November 1994. Alexandria, Va.: ACAA.
Behrens, G.P., and L.J. Holcombe. 1992. Boiler Chemical-Cleaning Waste Management Manual. EPRI Report TR-101095. Palo Alto, Calif.: Electric Power Research Institute.
Brendel, G.F., and T.N. Kyper. 1992. Institutional Constraints to Coal Fly Ash Use in Construction. EPRI Final Report TR-101686. Palo Alto, Calif.: Electric Power Research Institute.
Deloitte Touche Tomatsu International (DTTI). 1993. Coming Clean: Corporate Environmental Reporting. International Institute for Sustainable Development, and SustainAbility, Ltd. London: DTTI.
Electric Power Research Institute (EPRI). 1988. Fly Ash Construction Manual for Road and Site Applications. EPRI Report CS-5981. Palo Alto, Calif.: EPRI.

Electric Power Research Institute (EPRI). 1993. Options for Handling Noncombustion Wastes. 2nd ed. EPRI TR-103010. Palo Alto, Calif.: EPRI.
Electric Power Research Institute (EPRI). 1994. Management of Life Cycle Costs. EPRI Research Project E006. Palo Alto, Calif.: EPRI.
Elkington, J., and N. Robins. 1994. The Corporate Environmental Report: Measuring Industry's Progress Towards Sustainable Development. New York: United Nations Environmental Program.
EPRI Journal. 1992. Electricity for increasing energy efficiency. 17(3) (April/May).
Fiksell, J. 1993. Applying Design for Environment in the Electric Power Industry. Mountain View, Calif.: Decision Focus.
Foundry Management and Technology. 1991. May, p. 47.
Grann, H. 1997. The industrial symbiosis at Kalundborg, Denmark. Pp. 117–123 in The Industrial Green Game: Implications for Environmental Design and Management, D.J. Richards, ed. Washington, D.C.: National Academy Press.
Hedstrom, G.S., and R.A.N. McLean. 1993. Six imperatives for excellence in environmental management. Prism, 3rd Quarter.
Lott, T., L. Holcombe, and W. Micheletti. 1989. Boiler chemical cleaning waste treatment and disposal options. Paper presented at the 50th Annual Meeting of the International Water Conference, Pittsburgh, Pa., Oct. 23–25.
Metcalf and Eddy Inc. 1991. Wastewater Engineering, p. 166. New York: McGraw-Hill.
Patelunas, G. M. 1988. High Volume Fly Ash Utilization Projects in the United States and Canada. 2nd ed. EPRI Report CS-4446. Palo Alto, Calif.: Electric Power Research Institute.
Popoff, F., and D.T. Buzelli. 1993. Viewpoint: Full cost accounting. Prism, 3rd Quarter.
Sauber, B. 1992. Mobile demonstration plant will produce fly ash-based cellular concrete. Concrete Technology Today 13(1)(March).
Tibbs, H. 1992. Industrial ecology, an environmental agenda for industry. Whole Earth Review 77:4.
United Nations Environmental Program, Industry and Environment Office (UNEP/IEO). 1991. Companies' Organization and Public Communication on Environmental Issues, UNEP Industry and Environment Office Technical Report Series, No. 6. Paris: UNEP/IEO.
U.S. Environmental Protection Agency (EPA). 1994. Medical Waste Incinerators Background Information For Proposed Standards and Guidelines; Industry Profile Report For New and Existing Facilites. Washington, D.C.: EPA.
Wilson, J.S., and J.L. Greeno. 1993. Business and the environment: The shape of things to come. Prism, 3rd Quarter 1993.
World Commission on Environment and Development. 1987. Our Common Future. New York: Oxford University Press.
Yeager, K.E. 1994. Technology triggering structural change. Electric Perspectives 18(1).

The Pulp and Paper Industry

A. DOUGLAS ARMSTRONG, KEITH M. BENTLEY,
SERGIO F. GALEANO, ROBERT J. OLSZEWSKI,
GAIL A. SMITH, AND JONATHAN R. SMITH JR.

INTRODUCTION

Like other industries, the pulp and paper industry (referred to in the rest of this paper as "the industry") has come under increasing scrutiny for its potential environmental impacts. More than many other industries, however, this industry plays an important role in sustainable development because its chief raw material—wood fiber—is renewable. The industry provides an example of how a resource can be managed to provide a sustained supply to meet society's current and future needs.

This paper looks at the U.S. industry's current experience and practices in terms of environmental stewardship, regulatory and nonregulatory forces, life cycles of its processes and products, and corporate culture and organization. It describes near-term expectations in these areas and examines opportunities for overcoming barriers to improvement. It also provides an industry perspective on the most significant environmental issues of historical and future importance. Although the emphasis here is on complexity, shortcomings, and barriers, the industry has, in fact, continually improved its environmental performance while increasing its business. The problem areas are given more emphasis to highlight some of the challenges to be addressed.

ENVIRONMENTAL STEWARDSHIP

Wood is the chief raw material of the pulp and paper industry. In 1991, the worldwide harvest of roundwood for lumber and wood-panel products as well as pulp and paper was 1,599,272,000 cubic meters (Canadian Forest Service, 1993), or roughly 960 million metric tons. Approximately 63 percent of that wood, or 605 million metric tons, was used to manufacture 243 million metric tons of pulp, paper, and paperboard. The U.S. share of that production was 71 million metric tons, or 29 percent of the total, according to the 1994 Lockwood-Post's Directory of the Paper and Allied Trades (Miller Freeman, 1994). The industry in the United States employs over 690,000 people.

The availability and affordability of forest products and the economic health of the industry have always depended on the sustainable use of the forest resource. Significant management and technological improvements, such as plantation forestry and the development of the chemical-recovery cycle, were made a half-century ago. These improvements and improvements made since then have contributed to the sustainability of the industry and to the health of the environment.

The industry believes that its current industrial operations affect the environment minimally, due to the many improvements the industry has made to its environmental practices. However, as consumer and government concern about environmental impacts grows, the industry's environmental performance will be increasingly scrutinized. This scrutiny and the industry's commitment to improving its practices on the basis of good science and sound economics suggest possible changes in environmental practices on several fronts.

Silviculture: Managed Forestry

Intensive forestry, or silviculture, involves the efficient production of wood resources and has features in common with agriculture. Forestry, however, uses land far less intensively than agriculture, because the growth rotation cycles of trees require years, not months. Also, unlike most agricultural harvests, typically only a fraction of the growing forest is harvested in any given year. On a 15-year-average rotation, for example, one-fifteenth of the acreage on a tree plantation will be clear-cut in a given year. Thus, the majority of the land involved remains undisturbed, save for occasional clearing of understory to reduce competition for nutrients. Reforestation, the replanting of harvested acreage (or acreage lost to fire or flood), is another practice of intensive forestry. It ensures that the average rate of wood growth, expressed as the increase in the volume of wood per year per acre, is higher than it would be if the woodland were not harvested at all or were left to regenerate naturally.

As in agriculture, silvicultural practices include genetic improvement, regeneration, scientific management, appropriate scheduling of harvests, fertilization, and control of competing vegetation, insects, and disease. These practices help minimize the acreage needed to harvest a unit of wood. Silviculture, like agriculture, also must take into consideration the non-point-source pollution that could arise as a result of erosion and chemical applications such as fertilizing. Because silviculture is less intensive than agriculture, the risk of environmental damage is less than it is in the case of agriculture.

Reforestation results in more trees being planted, by a wide margin, than are harvested. There might be as many as six times more trees planted than harvested, depending on who owns the land (the ratio is higher on lands owned and managed by forest product companies) and whether or not the industry is at a higher or lower productivity level. Reforestation not only contributes to forest cultivation but also compensates for trees lost to fire, insect damage, and floods. Under current regulations, many smaller landowners in the United States might opt not to replant at all or might convert the land to agricultural or other use.

In addition to the measures outlined above, other improvements are being made on commercial forest lands, as appropriate. These include practices to enhance protection of various biotic species, wetlands, and water quality. In general, the basic practices of intensive forest management represent significant advancement in sustainable use of wood resources. Sustainable development, based on good science, is a goal that now guides the industry's practice. The principles of sustainable forestry (Box 1) adopted by members of the American Forest and Paper Association (AFPA) show the industry's commitment to the environment (American Forest and Paper Association, 1995).

Today, the industry in the United States gets the bulk of its raw material from nonindustrial private landowners. Intensity of harvesting varies significantly on these sites and depends on the objectives of the individual landowners. Private landowners get professional advice from a variety of sources, such as state forestry organizations, consultant foresters, and industrial landowner assistance programs. Some forest products companies provide support to their suppliers through management assistance programs (MAPs). At a minimum, these programs provide training desired by environmentally conscious landowners. At the buyer's discretion, compliance with the sustainable forestry management principles can be a criterion for continuing the business relationship.

The industry faces several forestry operation challenges. These include the harvesting methods used, the protection of threatened and endangered species, and potential restrictions on wood harvesting.

Harvesting Methods

There are two ways to determine length of the harvest cycle of trees: end use of wood or ecosystem impact. End use is the more direct method of determining the harvest cycle; dimensional lumber requires older trees than does pulpwood, for example. The effect of ecosystem considerations on the harvest cycle are more complex and might be driven by factors such as determining the proportion of older growth needed to protect a species habitat. Forest renewability includes practices such as clear-cutting that are perceived to be environmentally destructive. Depending on the degree and timing of commercial and environmental

> **BOX 1 Principles of Sustainable Forestry**
>
> - Broaden the practice of sustainable forestry by employing an array of scientifically, environmentally, and economically sound practices in the growth, harvest, and use of forests.
> - Promptly reforest harvested areas to ensure long-term forest productivity and conservation of forest resources.
> - Protect the water quality in streams, lakes, and other bodies of water by establishing riparian protection measures based on soil type, terrain, vegetation, and other applicable factors, and by using EPA-approved Best Management Practices in all forest management operations.
> - Enhance the quality of wildlife habitat by developing and implementing measures that promote habitat diversity and the conservation of plant and animal populations found in forest communities.
> - Minimize the visual impact by designing harvests to blend into the terrain, by restricting clear-cut size, or by using harvest methods, age classes, and judicious placement of harvest units to promote diversity in forest cover.
> - Manage company lands of ecological, geological, or historical significance in a manner that accounts for their special qualities.
> - Contribute to biodiversity by enhancing landscape diversity and providing an array of habitats.
> - Continue the prudent use of forest chemicals to improve forest health and growth while protecting employees, neighbors, the public, and sensitive areas, including stream courses and adjacent lands.
> - Broaden the practice of sustainable forestry by further involving nonindustrial landowners, loggers, consulting foresters, and company employees who are active in wood procurement and landowner assistance programs.
> - Publicly report progress in fulfilling commitment to sustainable forestry.
> - Provide opportunities for the public and the forestry community to participate in the commitment to sustainable forestry.

needs, many other, lower-intensity treatments such as periodic thinning or selective harvesting can be applied to industrial commercial forest lands. However, most North American commercial forests eventually need good sunlight to reproduce successfully, and clear-cutting is used to accomplish this goal. Moreover, plantation forests are usually more economically planted, managed, and harvested. Decisions on what harvesting techniques to use depend on the landowner's objectives, characteristics of the site, and forest conditions.

Protection of Threatened and Endangered Species

The forest products industry on its own and in collaboration with governments is intensifying efforts to protect threatened and endangered species. Companies are helping to identify the presence of such species on managed lands and are incorporating specific habitat management schemes into existing stand management and harvesting programs.

The species-by-species approach, inherent in efforts to protect threatened and endangered species, does not take into account the impact of intensive forestry on entire ecosystems and the many species living in them. However, in the absence of better understanding of the effect of harvesting and reforestation on ecosystems, it is the approach being used.

Potential Restrictions on Wood Harvesting

Loss of biodiversity, potential global warming, and deforestation are increasing pressure to retire woodlands, particularly forested wetlands and old-growth forests, from production and to restrict the harvesting of wood from public lands. Yet, abandoning the productive use of forests might not be the best course to address these concerns. For example, existing forestry practices have been shown to increase landscape diversity and carbon sequestration beyond that achievable in the absence of human activity. Plantation forestry supports habitats for endangered species. At a minimum, the objective is to maintain, not necessarily expand, such populations on forest lands. Trade-offs will be necessary in making these environmental decisions.

Turning Wood into Paper

Paper can be made from virtually any fibrous material, including cotton, sugar cane, and bamboo, but the vast majority is made from trees. The species and variety of trees used are important determinants of the type of paper produced. Some trees, such as pine, yield long fibers that are strong and absorbent (good for paper towels, for example), whereas others, such as hardwoods, are shorter and form a smoother surface (good for printing purposes). In the tree, wood fibers are bound together by an organic polymer called lignin. To make paper, individual wood fibers must be separated from each other (defiberized) in one of a variety of pulping processes. The separation can be achieved mechanically by grinding the lignin (groundwood process) or chemically by dissolving it. In chemi-mechanical or semichemical pulping, a combination of the two processes is employed.

Groundwood pulps, such as newsprint, are less costly to make but are of lesser quality in terms of strength and brightness. Chemical pulping processes, which are described more fully below, remove more lignin and yield essentially individual wood fibers, which can be converted into many products, from linerboard (the walls of corrugated boxes) to tissue paper or magazine stock. After pulping, chemical pulps are washed to remove and recycle the chemicals used and might be bleached white in a variety of bleaching sequences depending on the desired end product. Finally, the pulp is dried on one of several types of paper machines or pulp dryers. In such machines, most of the water in the pulp is first squeezed out by passing the wet pulp web through a press. Coatings or dyes might be added to the web. Then, the pulp is dried to less than 10-percent moisture by heated steel rolls or hot air. At the end of the machine, the dried pulp might be baled and sold as market pulp for converting to final products in other facilities, or it might be converted directly into paper or paperboard, depending on the thickness and weight of the sheet, by passing it between high-pressure rollers called calenders.

The Pulping Process

The following discussion centers on chemical pulping processes (Casey, 1983; Saltman, 1983), which supply more than two-thirds of the world's wood pulp. Chemical-recovery processes are used routinely in most chemical-pulping processes. First used a half-century ago, these processes now result in the recovery of 90 percent of the inorganics, which are reused as described more fully below. Nearly 100 percent of the dissolved organics are converted to energy.

The most widely used chemical pulping process in paper making is the sulfate, or kraft, process. It was invented in 1889. In the 1930s, the process was enhanced with chemical recovery. Other chemical-pulping processes (mainly using acid sulfite and soda) are sometimes combined with various chemical-recovery subprocesses. The typical kraft process involves turning logs into wood chips (Figure 1), which are then pulped (Figure 2).

The wood chips are pulped under high heat and pressure in continuous- or batch-digestion processes using white liquor (a water-based solution of sodium hydroxide and sodium sulfide). The white liquor dissolves the lignin and frees the cellulose fibers. Some of the cellulose is hydrolyzed to methanol, acetone, and other volatile and water-soluble organics. Some of the cellulose reacts with the sulfide ion to produce sulfonated organics, such as methanethiol, which can cause odor problems. When digestion is complete, the digester contains a mixture of brown stock (wood fiber) and black liquor. Black liquor is a mixture of sodium compounds (sodium hydroxide, sodium sulfide, sodium sulfate, and sodium carbonate), organic compounds, and salts including lignins and resins. Turpentine (a mixture of branched aromatic hydrocarbons) is also released from the wood in varying quantities, depending on the wood species.

Substances that are gaseous at digester pressure and temperature, including some methanol, acetone, organosulfurs, and most of the turpentine, are vented during digestion to condensers, where the turpentine is recovered for sale. Typically, nearly 100 percent of the noncondensible gases, which contain the odor

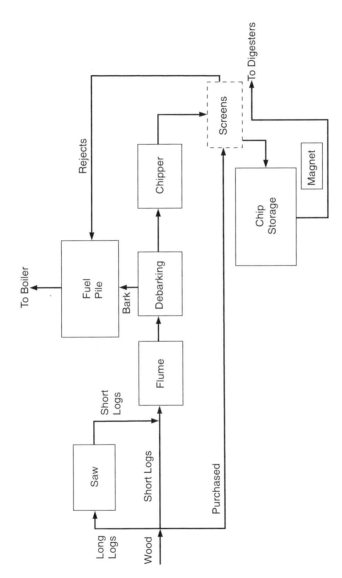

FIGURE 1   Wood-handling process.

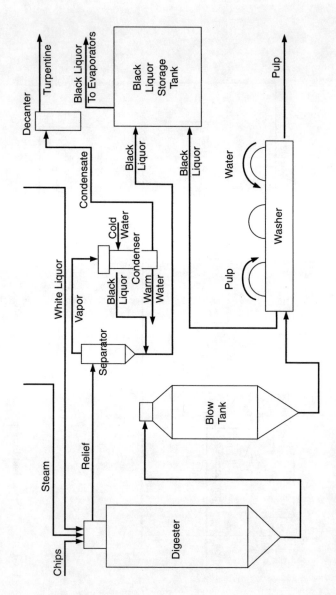

FIGURE 2 Pulping process.

compounds, methanol, and acetone are collected and destroyed by incineration in other combustion units in the mill. The sulfur dioxide from oxidation of the organosulfurs is generally scrubbed with alkaline liquors.

After digestion, the black liquor is separated from the brown stock pulp, usually by countercurrent drum washing. The brown stock might be screened and refined mechanically either before or after washing. Some methanol, acetone, and odor compounds might be volatilized and released during washing. Depending on regulatory requirements and aesthetic considerations, some mills capture and incinerate the gases that are released. The brown stock from the washing process is further delignified by bleaching, or sent to a paper machine. About 1.5 percent of the original weight of wood is dissolved organic material lost to waste treatment, and another 1 to 3 percent is fine fiber lost in primary and secondary wastewater treatment facilities. Typically, 90 percent of these losses are removed from the effluent prior to discharge to the environment. The resulting fiber fines and waste-treatment sludge—both of which are nontoxic and nonhazardous solid waste—constitute the majority of the solid waste generated by a pulp mill.

The black liquor, now containing about 7 percent inorganic salts and about 7 percent soluble organic material, is routed to evaporator systems, which increase the total solids to 50 to 75 percent to sustain combustion. During evaporation, additional methanol and odor compounds are evolved from the liquor. The vapor fraction is incinerated in the same combustor used to incinerate the digester noncondensibles. Before the mid-1970s, direct-contact evaporators (DCEs) were retrofitted on recovery furnaces to recover particulate matter from the hot flue gases, which, in turn, concentrated the solids to about 65 percent before firing. Later, black-liquor oxidation systems were installed to convert the sulfide content in the black liquor to stable materials to meet the total reduced-sulfur (TRS) emission regulations. However, newer low-odor furnace systems use additional evaporator units called concentrators instead of the DCEs. The net effect has been the removal of about 20 percent of mill odor emissions at an additional capital cost of several million dollars. Evaporator condensates are generally recycled for use as wash water for the pulp. To maximize water reuse, the more odorous of these condensates are often steam stripped and incinerated.

During the evaporation of black liquor from softwood pulping, tall oil soap (a mixture of sodium resinates named after the Swedish word for pine) floats to the surface of the liquor. It might be skimmed off and acidulated to the oil form. Rather than burn the soap along with the rest of the organic materials in the black liquor, it is sold to refiners for use in a variety of products from paper additives to cosmetics. Acidulation of soap liberates hydrogen sulfide, which is often collected by mills and scrubbed or incinerated.

The concentrated, or heavy black, liquor is fired in a specially designed chemical-recovery furnace, where the organic portion is combusted to produce steam and subsequently electrical energy by cogeneration. The inorganic portion, now separated from the organic portion, is recovered and converted back into the chemicals used for pulping. It forms a molten smelt in the bottom of the furnace, where sulfate is reduced to sulfide. This smelt runs off into dissolving tanks and results in green liquor—an aqueous solution of principally sodium sulfide and sodium carbonate. The boiler gases contain carbon dioxide, carbon monoxide, oxides of nitrogen, sulfur dioxide, traces of hydrogen sulfide, and sodium sulfate particulate. In most instances, a high-efficiency electrostatic precipitator is used to remove the particulate matter, which is redissolved in the incoming black liquor. Emissions of reduced sulfur compounds, carbon monoxide, oxides of nitrogen, and sulfur dioxide are controlled by proper furnace design and management of fuel and air. Hydrogen sulfide, or TRS, emissions are regulated by operational permit and are generally low enough to avoid causing ambient odor.

Dissolving the smelt liberates some hydrogen sulfide and particulate matter, which are controlled by alkaline scrubbers mounted on the dissolving-tank vents. A few mills also trap the odors given off by black-liquor storage tanks and incinerate them.

The green liquor from the dissolving tanks goes to a recausticizing system, so named because calcium oxide is added to it to convert the sodium carbonate to caustic soda. The green liquor is thus converted into white liquor, thereby completing the main chemical-recovery cycle. Inorganic impurities in the liquor cycle, such as silica and iron, are separated as dregs and grits. The dregs and grits are disposed of as solid waste, although some uses for them as inert filler have been found.

Another recovery cycle is employed to recycle the calcium used for recausticizing. The calcium carbonate formed in the causticizers is removed by gravity in white-liquor clarifiers or by filtration and washed to remove sodium salts. The weak-wash water, still highly alkaline, is generally recycled to the smelt dissolving tanks. The calcium carbonate "mud" is then thickened and introduced into a specially designed kiln, where it is calcined back to calcium oxide using an oil or gas flame. Bag filters are often used to control particulates arising from the handling of dry lime.

The lime kiln produces emissions that must be controlled. Particulates are removed by either a high-efficiency scrubber or an electrostatic precipitator. Traces of sulfide in the mud liberate hydrogen sulfide by carbon-dioxide stripping, so hydrogen-sulfide emissions are controlled by effective mud washing, scrubbing, and sometimes by oxidation.

In addition to the main chemical-recovery furnace, most kraft pulp mills burn wood waste associated with the pulping process in boilers. This waste is in the form of bark from pulp logs and chip fines and oversize material removed in the screening of wood chips that cannot be reused. Primary clarifier sludge might also be disposed of by burning. If the total steam produced from the recovered black-liquor solids in the chemical-recovery boiler and the normal quantity of wood waste in the wood-waste boiler are insufficient for the needs of the mill, additional wood waste might be purchased. The deficit might also be made up with fossil fuels such as coal, gas, or oil. Older kraft mills that use fossil fuels for energy might use a third boiler. These older mills generally have to use some fossil fuels because the older processes are less efficient. Newer kraft pulp mills derive almost all necessary steam and electrical energy from renewable resources such as black liquor and wood waste, and therefore require the use of very little, if any, fossil fuels. This, however, does not apply to lime kilns, which are generally fired with gas or oil.

Bleaching of Pulp

The further delignification, or bleaching, of pulp produces additional air and water discharges. Bleaching involves adding chemicals to wet pulp to remove more lignin (color). There is considerable variation in bleaching processes. Until recently, the preferred sequence to produce high-brightness pulps was the application of chlorine gas or chlorine water ("C" stage), followed by aqueous caustic soda and/or sodium hypochlorite extraction ("E/H"), followed by aqueous chlorine dioxide application ("D"), followed by another E stage and another D stage. The effluents from the alkaline E stages contain organic material equivalent to 1 to 3 percent of the total pulp weight as the result of delignification and some breakdown of cellulose. The effluent from the latter E stage is generally reused at least once for pulp dilution and washing in a "jump-stage" manner from E to E/H. The acid effluents from the D stages are reused in preceding acid stages in a similar jump-stage manner. Thereafter, they are generally handled separately to avoid evolution of hydrogen sulfide in the other mill effluents.

In recent years, concern has arisen over the formation of chlorinated organic compounds, including dioxin, during chlorine bleaching. The industry responded to the public perception of possible harm, choosing to take remedial actions even before major questions about the toxicity of these compounds were answered. Almost all mills now have taken steps to reduce the formation of dioxin and other chlorinated organics. Typical methods include reducing hypochlorite use, which reduces emission of chloroform, and substituting chlorine dioxide for chlorine in the C stage of bleaching, which reduces the formation of dioxin and other polychlorinated organics.

Solids dissolved in bleach-plant effluents have historically been pumped to external treatment prior to being discharged into the environment. Their recovery and reuse through the liquor chemical-recovery cycles have been rare because added evaporation requirements and the presence of chlorides, which are corrosive above certain concentrations in the process liquors, can lead to explosions in the smelt dissolving tanks.

Chlorine dioxide, substituted for chlorine to reduce the formation of chlorinated organic compounds, is an unstable compound and must be generated at the point of use. Proprietary methods for doing this involve the acidification of sodium chlorate in the presence of a reducing agent and sometimes a catalyst. The chlorine dioxide that is formed is removed by air or water vapor and dissolved in chilled water for storage. Although these systems are generally self-contained, emission collection systems are provided to collect and scrub any chlorine dioxide that might escape. Sodium sulfate (saltcake), a by-product of the acidification process, is sometimes recycled and used in the chemical-recovery loop. In mills that have such generators, the saltcake can offset some of the sodium and sulfur normally lost to air or water, as solid waste, or in product during the chemical-recovery process. This reduces the amount of sodium and/or sulfur that must be purchased and injected into the recovery cycle.

Chlorine and chlorine dioxide used in bleaching can potentially be emitted from the bleach reactor towers, the bleached-pulp washers, and associated filtrate-collection devices. Because excessive emissions are synonymous with excessive chemical use and cost, these emissions are controlled by continuously improving the control technology for the bleaching process. Improvements include automatic sensing of brightness and other parameters, and the use of algorithms to control the rate of application of bleaching chemicals to the pulp. More recently, most states have put in place additional regulations to control chlorine and chlorine dioxide emissions. To meet these regulations, emissions are forced through alkaline and reductive scrubbers that remove chlorine and chlorine dioxide with high efficiency. Chloroform is not removed by alkali scrubbing, but concentrations of the compound can be lowered by reducing the use of sodium hypochlorite bleach and increasing the substitution of chlorine dioxide for chlorine.

Because of concern about chlorinated organics and the reusability of bleach-plant effluent, research into totally chlorine free (TCF) bleaching processes has accelerated in recent years. (For more information on pulp bleaching, see Dence and Reeve, 1996.)

Meeting Environmental Challenges Head On

In turning wood into paper, the industry has met and addressed several environmental challenges, as described in this section.

Conserving Water and Treating Wastewater

The pulping industry has long been considered an intensive user of water. Early in this century, a typical mill used 60,000 gallons of water to make 1 ton of bleached paper. Today, new mills can produce 1 ton of bleached paper with less than 10,000 gallons of water. Process innovations, such as high-consistency bleaching and hot-stock screening, require less water. Noncontact cooling, which segregates water from contamination, has also reduced the quantity of water used. Additional savings have accrued from internally recycling water by using countercurrent washing and by reusing condensates, cooling and sealing waters, machine white water, and treated effluents. As a result, every gallon of water is reused an average of seven times within the process.

Of course, it takes capital to handle water and to pay for the energy to heat it and move it around. This has been a major factor in the development of new technologies. From a water-use perspective, there are few financial, legal, or physical reasons for the industry to lower water use in paper mills. In some situations, water conservation has been pursued to reduce the costs of waste treatment, which is largely a function of the volume of water treated. The main impetus behind conservation efforts to date, however, has been the general principle of environmental stewardship that less is better.

One of the key strengths of the U.S. pulp and paper industry has been its treatment of wastewater. For the past 30 years, long before they were required by law to do so, U.S. mills have been installing secondary biological treatment systems that reduce biological oxygen demand (BOD) and suspended solids. At locations where land area is available, aeration stabilization and activated sludge technology provide relatively low-cost ways to cut BOD by more than 80 percent. Oxygen-enriched technology became available in the early 1970s in time for use in space-limited facilities. Extensive study and optimization over the last 2 decades have raised the efficiency of these same plants to around 90 percent.

Solid Waste Disposal

Until the early 1980s, disposal of solid waste from pulp and paper production was not a problem. For one thing, the waste is considered nonhazardous by the Resource Conservation and Recovery Act (RCRA) of 1976. In addition, landfill rules required mainly the control of vermin and litter, access control, control of drainage, safe operation, and earthen cover. More recently, new rules on waste characterization, groundwater monitoring, lining and leachate control in some cases, and financial assurance have been added. It is difficult to get a permit for a new landfill today because of the public perception that there is something inherently dangerous about landfills. Therefore, there is an impending capacity shortage at many locations. Such local shortages are driving the exploration of technologies to reduce the volume of solid waste generated or to find new uses for waste.

Some mills, as part of their environmental programs, have developed systems to mix dried primary and secondary waste-treatment sludge with wood waste and bark for use as fuel. In most instances, they have carefully analyzed the waste and the combustion products to isolate potential problems with contaminants and have taken whatever steps were necessary to eliminate them. Nevertheless, opposition to incineration is often stronger than opposition to landfills. Yet, scientifically acceptable incineration systems are necessary for environmentally sound waste management in all industrial sectors. Better communication to the public by Environmental

Protection Agency (EPA) and industry representatives about the good science behind assessment of comparative risks is essential to breaking this deadlock.

Sludge can be used to amend soil, as landfill, as animal bedding, or to produce ethanol or other chemicals. Source reduction through such steps as improved retention of pulp fines can also help reduce waste. However, these and other potential applications are unlikely to negate altogether the need for new landfill capacity.

Controlling Spills and Leaks

An area of significant environmental stewardship in the industry has been in reducing spill and leak hazards. Most mills have already identified the tanks that have the greatest potential to cause injury or other damage in the event of leakage or rupture and have installed secondary containment, sumps, leakage detectors, and other appropriate safety measures. Likewise, leakage monitoring and alarm systems for compressed gases, especially chlorine, are now commonplace.

Conserving Energy

Energy conservation and the use of fossil-fuel alternatives play an indirect but significant role in environmental stewardship in the forest products industry. The standard practice of using bark and wood waste and black liquor as fuel eliminates about 54 percent of the demand for fossil fuel in the U.S. forest-products industry as a whole, including integrated pulp and paper mills (mills in which the papermaking operation is contiguous with the pulping operation) and nonintegrated mills (American Forest and Paper Association, 1994). Modern kraft pulp mill operations, with the exception of the lime kilns, can satisfy their total steam and electrical energy requirements using black liquor and wood waste and therefore do not require fossil fuels. Wood waste and black liquor are carbon neutral; that is, when burned they cause no net change in the carbon content of the biosphere over the harvest cycle and, therefore, do not contribute to the formation of so-called greenhouse gases. Other key energy conservation measures commonly used today involve reduced water usage; energy recycling and reclaiming in digester areas; systems to improve management and reclamation of low-level heat, for example, from recovery systems; and improved insulation.

The industry has always been interested in reducing the energy needed for pulping. Energy savings not only conserve fossil fuels, but also reduce emissions from the boilers used to generate steam or electricity. Despite its extensive use of self-generated energy, the industry is still the third largest industrial user of fossil fuels in the United States. Much of this energy is used to evaporate water—either from the black liquor prior to burning or from the paper after the sheet is formed on the machine. Processes are being developed and used that utilize low-pressure steam, which was previously vented, to evaporate water. One example is the steam that flashes off pulp when it is blown from a digester after cooking. This steam, called blow heat, can be routed to heat liquor in blow-heat evaporators. Low-pressure steam can sometimes be converted economically to high-pressure steam by vapor recompression.

On a paper machine, it is much less energy-intensive to remove water from a paper web by mechanical means than by evaporation. As much water as possible is squeezed out of the web in a paper machine press section, but the sheet is still 55 percent water. Extended nip pressing and impulse drying are novel mechanical dewatering methods that have emerged recently, and the future surely holds others. The general approach to energy reduction could include improvements in overall management of the reuse of low-level energy in both pulping and papermaking. Also in the pulping area, using nonfossil fuels in the lime kiln, or eliminating the kiln by developing autocausticizing processes, would contribute additionally to the fossil-fuel independence of the kraft process.

Recovered paper is also a great potential source of energy. Paper has significant heat value, burns cleanly, and is carbon neutral. Paper that is contaminated with materials (e.g., plastic) and difficult to recycle can be used as fuel to offset the use of fossil fuel. Because more locations are available for burning recovered paper than for reprocessing it, collection logistics can be simplified and transportation energy requirements reduced.

Controlling Odor

One of the ironies of the industry's attempt at environmental stewardship relates to odor control. It seems that when odor control equipment is installed at a kraft mill, the number of complaints about bad odors increases. People become accustomed to an odor when it is always around but notice it when it comes and goes. Today, with heightened environmental consciousness, people associate the smell of an industrial operation with exposure to a hazard. Hence, there will be increasing pressure to achieve and maintain odor control, possibly in excess of what is required by law. Technology offering superior odor control will be at a premium. The need for low-cost backup odor control devices is also apt to grow in the industry, whether or not they are required.

CURRENT AND FUTURE TRENDS SHAPING TECHNOLOGY IN THE INDUSTRY

Restrictions on the supply of wood will increase its cost. Cost increases, in turn, will make recycled product and process technology relatively more attractive. They will also raise interest in improving pulping processes to conserve cellulose and increase yield through catalytic pulping with anthraquinone, by defiberizing with less delignification through solvent pulping, or using new mechanical pulping technologies. In addition to having less desirable end-product characteristics, the mechanical processes do not compare well with the kraft process because of their large requirements for fossil fuels and production of waste streams for which commercial treatment technology has only recently been developed. Solvent pulping systems require more development to resolve strategies for using by-product carbon and to improve their overall economic feasibility. Comparisons between technologies need to be done on the total-mill concept (i.e., by looking at how each technology affects the economics, safety and environmental effects, and material and energy balances in all parts of the mill).

Of course, all technologies designed to improve the resource and economic efficiency of pulping and papermaking must pass environmental muster. Most of the first part of this paper has described the strides that have been made already by the industry to control pollutants. The remainder of this section describes new developments and trends that might affect the industry's environmental stewardship in the future.

Minimal-Impact Philosophy

The cornerstone of the minimal-impact philosophy is the effluent-closed mill. The common wisdom is that effluent closure requires eliminating chlorine from the bleach plant because of the corrosivity of aqueous chloride. This has led to the development of closed kraft TCF bleaching, which is not a field-proven process and might cause contaminants to build up over time. "Even with the elimination of chlorine, methods for controlling or purging the build-up of other materials, anionic trash for example, need to be found and improved upon" (Technical Association of the Pulp and Paper Industry, 1992). The conventional wisdom about the need to eliminate chlorine might not be achieved by any mill through effluent closure. Contaminants will still have to be managed as solid waste, as emissions into the atmosphere, or with the outgoing product. It might be possible to manage chloride levels of corrosion rates with alternative technologies. Separate chemical-recovery systems specifically for handling bleach-plant chemicals could be developed (Folke, 1994).

Regardless of the methods developed, the concept of closure can be deceptively appealing. It appears to minimize water use, lowers energy consumption, and lowers waste impact. Ozone, peroxide, and enzymatic bleaching systems are alternative processes under development to minimize concerns about chlorinated organic trace by-products, such as dioxin. However, little is known about the composition of the effluent from these newer processes or their potential for environmental or human health impacts. More study is needed on ways to trace chlorinated organics or reduce their generation in wastewater or in process streams. Some approaches that have been considered are biological or chemical dechlorination, wet oxidation, ultraviolet irradiation, and electrolytic treatment, as well as methods utilizing natural processes such as spray irrigation, overland flow, and use of either natural or man-made wetlands.

What is the industry likely to do with conventional emissions and discharges in the future? As far as water is concerned, aesthetics are likely to receive greater emphasis; in fact, the minimal-impact mill will be expected to have little noticeable environmental impact as far as the public is concerned. There is likely to be an increasing number of limitations placed on incremental allowable stream color, especially at locations near recreational areas. Therefore, research into lower-cost color removal techniques will continue.

Paper Recycling

Concern about deforestation (and public misperception that the forest products industry is a contributor to it) and about the worldwide production of solid waste have put paper at the top of the list for recycling efforts in developed countries. Of course, internal (intrafirm) recycling has long been practiced. Indeed, every paper mill has a repulper to handle its own waste. Now, the key interest is in post-consumer recycling. There is currently a glut of recycled fiber, because the capacity to collect recovered paper is greater than the capacity or technology to process it into products of any given level of quality, which in turn exceeds the market demand for most recycle-grade paper. In addition, the cost and energy requirements to process recycled fiber into high-quality grades exceed those for processing virgin timber. Another important factor is the cost of collecting and transporting recyclables to mills equipped to process them.

The U.S. pulp and paper industry and its customers are facilitating the development of recycling technology and capacity. They are doing this through a voluntary wastepaper recovery target (an effort of members of the AFPA) and through purchasing specifications for minimum recycle content on the part of many customers, most notably the federal government. The industry, through AFPA, committed itself to recovery for recycle and reuse 40 percent of all paper used by 1995. This goal was met 2 years ahead of schedule, and the industry has recently established a new goal: 50 percent recovery by 2000. The Technical Association of the Pulp and Paper Industry (1992) projected that meeting the year 2000 goal would essentially require a doubling of the 1985 recycle capacity by 1995, assuming no change in processing technology. The principal technologies used by the paper recycling industry today are limited to classification at the point of collection, limited plastics and metal removal at the point of remanufacture, and surfactant deinking. Management of the solid waste generated as a result of recycling efforts presents a significant technical challenge. Research to explore a variety of alternative uses for this waste, including as fuel, compost, and feedstock for ethanol production, is ongoing.

The expansion of paper recycling will increase interest in deinking and rebleaching technology, used to remove color from the fiber, and in lowering the cost of doing so. Improvement will also be needed in the separation and disposal of deink sludge. Innovation in ink and copier-toner technology also holds promise for reducing the toxicity and amount of deink sludge.

It is widely believed that the trend in the pulp and paper industry is to move away from constructing virgin pulp and paper mills in "greenfields" (new, untouched locations) toward locating recycle-intensive "mini-mills" closer to urban population centers, where the recyclables are collected. If the future unfolds this way, disposal of the large amounts of deink sludge will have to compete with municipal waste for space in landfills. Finding new uses for this sludge is critical. To date, deink sludges, which are made up primarily of short fibers and clays, have proved to be superior landfill cover materials, adsorbents, and aggregates. New chemical and mechanical operations to deal effectively with the separation of useful materials from contaminants in recovered papers are urgently needed. A consequence of the development of urban mini-mills will be the reduction in recycle fiber available for use in remote, larger paper mills. A national standard for recycle content, therefore, might be difficult to achieve.

Other Factors

The development of new or improved technologies for pulping, bleaching, papermaking, and chemical recovery can potentially further improve the efficiency of resource utilization in the pulp and paper industry. Alternative bleaching processes that reuse bleach plant wastewater would reduce water use. The development of black-liquor gasification technology could improve the safety and efficiency of the chemical-recovery process and reduce emissions of air pollutants. Finding alternative uses for the solid waste resulting from increasing production of recycled paper (e.g., deink sludge) will also help reduce waste.

However, good environmental practices must be based on science. They must proceed with the same deliberation and objectivity as any other technical endeavor. Efforts to find new uses for waste materials need to take into account other resources used. Even alternatives that are less acutely toxic could cause greater environmental damage, in terms of wasted energy or the production of a larger amount of less-toxic pollutants. Finding new uses for waste and substituting less-toxic materials for ones that are more toxic make sense as general principles but should not be interpreted as inviolable rules.

In undertaking major capital projects, companies in the industry minimize the possibility of not addressing critical environmental concerns by having the projects reviewed within the organization by a group responsible for environmental oversight. Life-cycle assessments, discussed later in this paper, could guide internal corporate decision making on environmental issues. However, there is much work to be done, such as getting the data necessary to define boundary conditions for reliable life-cycle assessments.

REGULATIONS AND STANDARDS

The inefficiency of approaching environmental issues in a piecemeal manner is glaringly evident from the last 2 decades of environmental initiatives. At times, money was spent on technologies that later were found to be inadequate. Sometimes, solutions created other, unforeseen problems. At other times, the opportunity to solve multiple problems simultaneously with a single but different technology was missed. Some companies, using long time horizons, are attempting to accurately predict what entire plants will need in terms of process and environmental control for the next 20 years or more. Even this more methodical approach is problematic, however, because regulatory requirements will change in response to changing science, economics, and public demand. Creating no environmental impact is technically impossible. Most human activities alter the environment in some way. "Minimal impact" might be a realistic goal, although its definition is vague. Indeed, this vagueness offers the industry an opportunity to work with various stakeholders to reach common ground on which to build sound environmental policy and strategies.

Regulators, industry, and other stakeholders should work in close cooperation to set long-term environmental goals. Such cooperation is currently lacking in the U.S. regulatory efforts. Allowing the industry to experiment with alternative means of meeting these goals would be a positive departure from the current command-and-control approach.

Regulatory Trends: Where We Are Today

Prior to the creation of EPA by Congress in 1970, the pulp and paper industry in America operated under an assortment of state regulations of varying stringency. As early as the end of World War II, industry environmental performance was well ahead of government initiatives. The National Council for Air and Stream Improvement of the U.S. pulp and paper industry began disseminating environmental technology to its members in the 1940s. By the 1960s, most new mills were built with both primary and secondary wastewater treatment, as well as electrostatic precipitators on boiler stacks and scrubbers on lime kilns and smelt tank stacks, even though they were not required by law in most localities. Since the 1970s, however, federal regulations, administered by EPA and individual state authorities, have grown exponentially (Figure 3), becoming the dominant force shaping the industry's response to environmental issues.

Each decade has heralded a new regulatory wave. The 1970s might be thought of as the "framework decade," in which the basic federal structures of delegation to the states and permit-granting systems were laid out and tested. The Clean Air Act of 1970, for example, set limits on ambient air contamination and required implementation plans from the states, including permit-granting rules and programs for solving ambient-air-quality problems. The Clean Water Act of 1972 established the National Pollutant Discharge Elimination System, which required every manufacturing and municipal waste-treatment facility to obtain a permit limiting discharges. The Endangered Species Act of 1973 mandated protection of threatened and endangered species via the U.S. Fish and

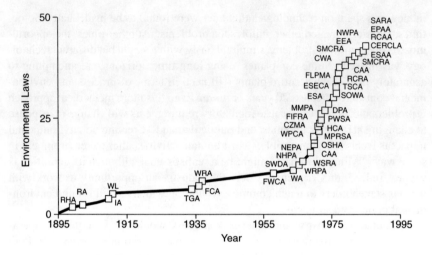

FIGURE 3   Growth in the number of U.S. environmental laws. SOURCE: Balzhiser, 1989.

Wildlife Service's rulemaking process. The Resource Conservation and Recovery Act of 1976 brought solid and hazardous waste under uniform federal regulation for the first time, and the Toxic Substances Control Act of 1976 addressed use of chemicals.

The 1980s might be considered the "land decade." It was during this period that federal authority expanded beyond protection of ambient air and water. The

Comprehensive Environmental Response, Compensation, and Liability Act of 1980, popularly known as Superfund, and the Superfund Amendments and Reauthorization Act of 1986, went far beyond the original scope of the Resource Conservation and Recovery Act, covering accidental releases, leakage from underground tanks, and remediation of disposal sites. There was also a shift away from the traditional regulatory process of involving solely technical professionals in the industry and government toward greater citizen involvement in permit hearings and lawsuits. The 1980s were also characterized by increased pubic awareness of environmental issues.

The 1990s began with reauthorization of the Clean Air Act, which greatly expanded the list of regulated air pollutants to include toxic compounds. An upshot of this has been concern that air controls are merely transferring pollution from air to water and soil. This has led to the so-called cluster rules and multimedia limitations now under development. Other regulatory and voluntary initiatives, spurred by concern over global environmental effects, are now directed increasingly at pollution prevention, reducing the discharge of pollutants in general, and at a long list of individual substances, including carbon dioxide. It looks as if this will be the pattern for the rest of the decade, and it seems apt to dub it the "prevention decade."

The following is a summary of the environment-related regulations affecting the pulp and paper industry in the United States today:

- *Air.* Stack emissions of particulate matter, sulfur oxides, nitrogen oxides, carbon monoxide, volatile organic compounds, and TRS are usually regulated by renewable permit. Numerical permit limits are the more stringent of either technology-based or air-quality-based limits. Technology-based limits derive from the typical performance of some type of control, such as best available control technology, whereas air-quality-based limits derive from the maintenance or improvement of the quality of the ambient air or the prevention of significant risk to health. Standards also exist for odors and effects on visibility. Intentional discharge of chlorofluorocarbons into the atmosphere is prohibited. In 1995, EPA finalized maximum-achievable control technology (MACT) standards, which established limits for a number of hazardous air pollutants. According to the current schedule, chemical pulp mills will be required to comply with these limits by 1998. The combination of these standards with pending revisions to effluent guidelines is referred to as the cluster rule.
- *Water.* Discharges of conventional pollutants, such as BOD, suspended solids, corrosives, oil and grease, fecal coliform, and 127 pollutants designated as toxics, are open to regulation through permit renewals, as are discharges of other nonconventional pollutants, such as color or chemical oxygen demand and stormwater outfalls. Permit limits are either technology based (mass allowed per unit of product) or water-quality based (not allowed to exceed or worsen noncompliance with federal or state water-

quality criteria, whichever are more stringent). Water-quality criteria limit exposure of people and aquatic life to harmful constituents. Discharge of oil in any visible amount is prohibited. Effluent guidelines for the pulp and paper industry are currently being reviewed by EPA as part of the cluster-rule process. These standards will substantially increase the stringency and scope of existing technology-based standards as well as establish numerical limits for several compounds, including 12 chlorinated organics not previously regulated.

- *Solid and hazardous waste.* Most solid-waste disposal sites must obtain permits and follow specific management procedures. Federal standards exist for municipal waste disposal. Currently, industrial-process waste is regulated by the states. Hazardous waste, which is listed as hazardous according to either its chemical composition or any of four characteristics (toxicity, reactivity, corrosivity, ignitability), must be handled specially and treated before disposal. Leachate from new waste disposal sites must be collected and treated, if necessary, before release. Underground storage tanks must be tested for leakage, and any leakage must be remediated. Groundwater that might be affected by waste disposal or underground storage must be monitored and remediated, if necessary, to meet certain criteria. Old disposal sites that EPA determines pose a threat to human health or the environment must be cleaned up.
- *Protection of ecosystems.* Recent stringent interpretation of the Endangered Species Act and the Clean Water Act has come in conflict with traditional private-property rights. Controversies have resulted over such issues as protection of endangered species (e.g., the northern spotted owl in the Northwest) and use of property purchased before it was designated as wetland. The interpretation of these regulations and how they affect the forest resource is of concern to the industry today.
- *Health and safety of products.* Food packaging is regulated by the U.S. Food and Drug Administration and the U.S. Department of Agriculture (USDA) because of the potential for migration of substances from packaging into food. Food-packaging rules are contained in the *Code of Federal Regulations* (CFR), Title 21, mostly under Sections 174 and 176. USDA also regulates packaging components that could come into contact with meat or poultry. The agency relies on provisions in 21 CFR plus a certification system for suppliers of packaging.

Regulations also require that equipment containing polychlorinated biphenyls (PCBs) not be located in a plant such that leakage could contaminate the product. They also require that the concentration of PCB in the product not exceed 10 ppm. These regulations, in effect since the 1970s, are not an issue any longer for either virgin or recycled papers. Twenty years of monitoring indicates a clear overall reduction of PCBs to levels far below 10 ppm.

The Industry's Record

The industry considers its record of compliance with U.S. regulations to be excellent, especially considering the volume and complexity of the rules it is subject to. Many infractions are administrative in nature (e.g., failure to file the proper notifications or forms) or result from poor communication and follow-up on environmental needs. Nevertheless, enforcement of such infractions is being pursued with increasing vigor, so the record could wrongly imply declining performance. Many companies have set up environmental management systems that include environmental training for nonenvironmental professionals and environmental auditing to reduce the opportunity for infractions of this kind.

A partial list of the industry's environmental accomplishments is provided below (American Forest and Paper Association, 1994). The industry

- spent over $1 billion per year in the 1990s on pure environmentally related capital, expenditures that in 1991 and 1992 represented almost 20 percent of total capital expenditures;
- reduced in-mill water usage and the volume of effluent generated in the production of a ton of paper by over 70 percent since 1959;
- decreased the amount of BOD in effluent from 94 pounds per ton of product in 1959 to 8.2 pounds per ton in 1988;
- lowered the amount of total suspended solids discharged per ton of product by 80 percent between 1979 and 1988;
- reduced the proportion of sludge going to landfills and lagoons by 16 percent between 1979 and 1988;
- cut discharges of dioxin by 90 percent between 1988 and 1992, limiting the combined generation of the compound by all U.S. mills to less than 4 ounces per year;
- performed widespread process changes sufficient to decrease the amount of chlorinated organic compounds in effluent by 30 percent since 1988;
- cut use of elemental chlorine by 34 percent since 1986, with expectation of a 50-percent reduction by 1996;
- planned to increase dramatically the use of alternative bleaching chemicals, including chlorine dioxide and hydrogen peroxide (a 60-percent projected increase by 1996) and oxygen (a projected 90-percent increase);
- decreased releases of TRS by 73 percent between 1968 and 1980, as kraft pulping capacity was increasing 36 percent (By 1990, the reduction was estimated to be over 90 percent, despite a 75-percent increase in pulping capacity.);
- lowered total emissions of sulfur dioxide by 30 percent in the 1980s as the volume of paper manufactured increased by the same percentage, resulting in a 50-percent reduction in emissions per unit of production;
- was three times more successful at lowering sulfur dioxide emissions than U.S. industry as a whole;

- curbed growth of mill boiler fuel consumption sufficiently to produce a 15-percent decrease in the per-unit-of-production release of oxides of nitrogen during the 1980s;
- steadily lowered per-unit releases of acetone, chlorine, and chlorine dioxide into the air between 1988 and 1991, as chemical wood production increased 8 percent; and
- dropped per-unit emissions of chloroform into the air by 30 percent between 1989 and 1991.

Regulatory Trends: Where We Are Headed

On the U.S. regulatory scene, the greatest impetus will be to clarify remaining issues regarding requirements under the Clean Air Act Amendments (CAAA) of 1990 and integrate them with the pending cluster rule and, eventually, with the reauthorized Clean Water Act.

Perhaps of greatest concern to U.S. industry in the near future are methanol and chloroform emissions, two of the hazardous air pollutants (HAPs) identified for regulation under the CAAA. The emissions of these compounds are in decline. Chloroform emissions are reduced by using molecular chlorine and hypochlorite in bleaching; methanol is reduced by capturing it, along with noncondensible gases from pulping, liquor evaporation, and pulp washing, and then incinerating it. Most mills in the United States already have made changes to achieve reductions. If MACT is defined as the emissions level achieved by the best 12 percent of the industry, most U.S. mills will have to go further to meet the new standards.

The CAAA mandates that the residual health risk from HAPs, after MACT has been applied, be analyzed to see whether further reductions are needed to protect public health. Sources of HAPs will have to be further reduced depending on what level of risk is considered acceptable.

Risk assessment is often used by U.S. agencies to deal with specific environmental problems such as residual risk. Risk assessment, however, has not been adequately used by the government to prioritize environmental concerns. Meanwhile, the list of risk-based concerns continues to grow, and with it the enormous burden of doing proper risk assessment.

Concerns about regional air pollution have rekindled interest in two of the five criteria pollutants: sulfur oxides (and their contribution to acid deposition) and nitrogen oxides (and their contribution to regional levels of photochemical oxidants). Studies by EPA over the past 5 years suggest that nitrogen oxide emissions might play a greater role in the formation of photochemical oxidants than previously thought. The upshot of this finding is that many small boiler operations in regions that do not attain the current ambient-air-quality standard for ozone might have to install reasonably available control technology for reducing nitrogen oxides emissions. Installing such technologies, such as low-nitrogen-oxide burners or steam injection, will add costs due to the need for retrofitting and from associated losses in energy efficiency.

The trend today is for regulators to limit the choice of manufacturing process as a way of reducing pollutant emissions to all media. This is a shift away from the practice of attempting to control emissions to specific media at the end of the pipe. This trend is driven by the concern that the requirements of the CAAA might cause emissions of certain chemical compounds to be shifted from air into water or, as solid waste, into soil. In December 1993, EPA proposed a cluster rule that in effect limits the companies' ability to use process pulp bleaching. The pulp and paper industry was the first industry targeted for such a rule, which combines the requirements of the MACT standards with revised effluent guidelines under the Clean Water Act. Although EPA has for many years issued technology-forcing regulations, the technology being forced was for end-of-pipe pollution control. The new rules are the most significant effort undertaken by U.S. environmental regulators to force the type of manufacturing technology used. The cluster rule is exemplified by standards for all bleached-paper-grade kraft and soda operations based on the use of either oxygen delignification or extended delignification, and complete substitution of molecular chlorine with chlorine dioxide.

The Current Regulatory Process

The current regulatory approach is inherent in legal mandates passed on to EPA by Congress, which too often bends to environmental fashion rather than using scientific evidence to set a course of action for the long term. Constant jumping from one crisis to another compromises the success of existing programs, leads to regulatory overlap and duplication of effort, and confounds capital planning by affected enterprises. The current cluster-rule effort in the United States is an example of a belated and ad hoc attempt to resolve such problems, but it only treats one case, and in a crisis atmosphere (under court order) at that. There is every reason to expect more of the same unless certain barriers are overcome, as discussed more fully in the last section of this paper.

Litigation

A quasiregulatory force affecting all U.S. industries today is the threat of citizen-action lawsuits based on various environmental statutes or on allegations of injury. This activity will increase not only because of heightened media attention to the environment, but also because of the ever-increasing ability of science to relate cause and effect. The desire to prevent real harm to anyone and avoid the cost of defending against false allegations will drive companies to tread ever more carefully in environmental matters.

NONREGULATORY ENVIRONMENTAL PRESSURES

In addition to these expected regulatory activities, three relatively new nonregulatory forces are increasingly driving environmental effort and expenditures in the industry. The most significant of the three is the marketplace itself. This "green market," already heavily influential in Europe, will increasingly favor products that are biodegradable, low in toxicity, and reusable. Most of all, it will favor recyclability and high-recycled content.

Late in 1993, President Clinton ordered the executive branch of the federal government (i.e., nearly all federal offices) to specify a minimum amount of aftermarket recycled content in all paper it purchases (Executive Order 12873. Oct. 20, 1993). The requirement for recycled-fiber contents starts at 20 percent in 1995 and increases to 30 percent by 1999. Other major users of paper might adopt similar criteria.

As recycling and the capacity to process recycled material grow, new environmental demands will arise. The obvious one is that the amount of deink sludge will increase. In addition, new recycling mills are likely to be located closer to population centers. Thus, the impact of these operations on air and surface waters could add to environmental burdens in these areas, and mills will have to account for these potential impacts. Another adverse aspect of recycle-fiber processing is that it does not supply its own energy in the form of biomass, as does virgin-fiber processing.

Some believe that reducing environmental impacts is synonymous with economic savings and that environmental improvement would happen by itself if industrialists took a closer look at recycling and conservation options. Undoubtedly, there are some unrealized opportunities here, but it is easy to overestimate the degree to which such incentives can substitute for regulations. After all, it has always been the business of engineers to look continually for ways to increase net profit.

A broader version of this principle is embodied in the concept of full-cost accounting, which adds aesthetic, ecological, depletion, and contamination values to the cost of a resource. There is no consensus today about how these values would be established or how the accounting would work. The only way full-cost accounting per se can drive environmental improvement is if it results in increased profit. Unlike conventional costs, which are based solely on labor and material supply and are therefore self-determining, these new values would have to be established and funded by some national or international regulatory authority. To address regional and local environmental issues, one would have to either have local variations in values or create a hybrid with the current regulatory scheme. In any case, no one can say such a system would automatically be less contentious or complex than the existing framework.

A significant second force is increasingly stringent environmental regulations in countries around the world. Some countries limit the environmental effects not only of their own products, but also of imports. Such regulations might also take the form of limitations on the amount of pollutants emitted or discharged per unit of product at the point of manufacture. They might require ecolabels, which are then used as trade barriers.

The third force, which will become manifest in 2 or 3 years, is international environmental standards, which are currently being developed by the International Standards Organization (ISO).

For all of these reasons, unless there is a downward shift in the priority of environmental issues relative to other public concerns, the industry should expect equally or even more complex rules, ever-tighter regulatory restrictions, and a high level of competition in the green market. This will lead, in turn, to periodic surges in capital costs. Environmental operating costs should increase accordingly, and the industry can expect these costs to consume a larger portion of its resources in the years ahead.

ENVIRONMENTAL PERFORMANCE METRICS

The stringency of environmental regulations is increasingly linked to the measurement used. In the early 1970s, environmental measurements by companies consisted almost entirely of spot tests of emissions and discharges such as semiannual stack tests and daily effluent BOD, with a smattering of required ambient tests (e.g., river oxygen and ambient particulates). Regulators assumed that the spot tests were sufficiently representative of continuous performance. Independent verification or auditing of results was reserved for enforcement cases. Knowledge of the condition of the ambient environment was left up to the state and federal governments. Individual substances that either could not be detected by existing methods or were not believed to be present were not regulated. This relatively simple approach has changed dramatically over the last 20 years.

Emission and discharge monitoring technologies have become very sophisticated. As continuous and automated methods for determining pollutant concentrations in the environment and in organisms become more reliable, they are required in permits. Most significantly, as measurement sensitivity improves, substances can be found at lower concentrations. As a result, substances are found in effluents and emissions that had not previously been known to be present. If such a substance is believed environmentally significant, it is added to the lists of regulated substances. Perhaps the best-known example of this is dioxin, which can now be measured at concentrations as low as one part per quadrillion.

Mathematical modeling is used increasingly to estimate exposure to environmental contaminants. The movement of pollutants from sources to receptors has been difficult to measure because it is time dependent and because the ambient environment is vast and variable. Because of this, there has always been a gap between the concentration of pollutants predicted to exist in exposed organisms and the actual amounts found. There is therefore a need to improve mathematical modeling of the environment, for example by macromodeling media to determine exposure to pollutants; micromodeling the assimilation and metabolism of pollutants by organisms; and verifying model performance using measurement data.

The quality of assessments of exposure to pollutants, as determined by emissions testing and modeling, and of assessments of impacts on health, as determined by epidemiological study and pharmacokinetics research, are gradually converging. The overall result is an increased ability to relate an industrial activity to a particular health or ecological effect.

The net effect of these advances on the industry has been refinements in discharge limitations that are based on impact (as opposed to those based on available technology). It is now possible to calculate (with uncertain accuracy) a maximum-allowable rate of discharge of a pollutant using a defined maximum-allowable concentration of the pollutant in an organism. Thus, the measurement (e.g., testing for pollutants in fish) and limitation of pollutant levels in organisms constitute an increasingly common feature of permits. Many bleached-pulp mills, for example, test fish regularly and frequently for dioxin. State regulators then compare the results of these tests with the allowed maximum levels and decide whether to set permit limits.

Among the public, terms such as total, eliminate, and zero are often used in relation to environmental effects with no appreciation of the fact that they are meaningless outside the field of pure mathematics. As emissions and concentrations of contaminants and observable effects approach the limits of detection, it becomes difficult to distinguish real improvement from measurement "noise." This is important as it relates to the above-mentioned need for better setting of priorities.

Environmental performance is not measured solely in the ambient environment, of course, but is also indicated by business practices. As a way of proactively improving environmental performance, some firms are defining numeric parameters that measure gross compliance with the law and company policy (auditing), gross effect of specific products and their manufacture on the environment (life-cycle assessment and ecolabeling of products), and the effectiveness of environmental management systems. None of these approaches is yet the subject of regulation in the United States, even though many domestic, foreign, and international organizations are participating in programs to define and use such parameters to effect improvement in environmental performance and public relations.

Elements of these approaches already appear in voluntary programs such as CERES (Coalition for Environmentally Responsible Economies), PERI (Public Environmental Reporting Initiative), and the Japanese Keidanren, and they might eventually appear in the ISO international environmental standards.

Life-Cycle Assessment

The concept of product life cycle recognizes that the environmental impacts of a product extend beyond the product's disposal when its useful life is over. A clear profile of the environmental impacts of a product can be obtained by carrying out a life-cycle assessment (LCA). The assessment consists of a life-cycle inventory (a quantification of energy and raw materials requirements, and environmental releases throughout the life cycle of the product), a life-cycle impact analysis (a characterization and assessment of the effects of environmental loadings identified in the inventory), and a life-cycle improvement analysis (a systems evaluation of the needs and opportunities to reduce environmental impacts).

The life-cycle concept is not new to the forest products industry. A renewable raw material resource seems intuitively superior to the alternatives. The same is true for a resource that also provides for the self-generation of a majority of the industry's process energy. These benefits were understood long before life-cycle terms were applied to them. Now, it is important that the quantitative methods being developed for LCA accurately reflect these unique circumstances of the industry.

Life-cycle inventory studies were conducted originally in the United States and in Europe as an outgrowth of the oil crisis of the early 1970s. As the energy crisis faded, interest in the life-cycle concept waned. It was not until the mid-1980s that life-cycle inventory approaches reemerged. In the 1990s, LCA expanded rapidly as a result of a converging set of technical developments, business interests, and public policy and information needs. Directives by the European Commission's Directorate of the Environment concerning packaging, liquid-food containers, ecolabeling, and so forth played a key role in spurring international interest. The impending use of LCA as a market-based regulatory tool by this key economic bloc has piqued worldwide interest.

Interest is especially high in competing industry sectors, such as plastics and paper, in which the relative outcome of an LCA depends heavily on which factors are included in the inventory of effects and on the methodology used for the LCA itself. The marketing establishments of competing industries, especially those in consumer products, see LCA and similar ecoprofiles as tools to integrate environmental science into marketing efforts. On the plus side, such tools permit technically based product differentiation by environmental attributes. Having numerical results that lead to a ranking system for comparison of products is very appealing to marketing interests in general. Obviously, however, there is the potential for the LCA to be misapplied to secure market advantage, which underscores the need for a consistent and technically sound approach.

There is a gradually emerging consensus in some developed countries on what should be included in a life-cycle inventory. As of this writing, however, no complete LCA methodology is known to have been officially adopted in any jurisdiction, either by an agency of government or by a standards-making or certifying body. Nevertheless, the world is moving in that direction, and there is much international activity along those lines.

Since 1990, an American and European group, the Society of Environmental Toxicology and Chemistry (SETAC), has played a major role in the development of life-cycle applications. SETAC's efforts have helped define the limits of what can be done in LCA and have been picked up and further developed by many other environmental organizations, such as EPA and ISO.

Recognizing the inadequacy but growing importance of LCAs, AFPA is sponsoring a life-cycle inventory study in the grocery-bag sector to reflect intrinsic characteristics of paper products and, with NCASI, to enhance the present LCA methodology. The results should help overcome difficulties and delays in reaching agreement internationally on a proper and sound impact-assessment methodology.

Another objective of the AFPA study is to compare in an unbiased manner various attributes of paper bags with those of plastic bags. For this purpose, a peer review group of experts and stakeholders from both industries was formed and has been reviewing the study from its inception. This review group is a model of transparency in design and purpose. So far, the group has accepted the study's objectives, boundaries, scope, and database. A goal of this effort is to address a key inadequacy of LCA: The method does not link discharge inventories to actual environmental impacts. Until this shortcoming is remedied, LCA might actually serve as a barrier to real environmental improvements.

THE CHANGING CORPORATE CULTURE

Environmental concerns are currently a major issue for the pulp and paper industry, and they are expected to remain so far into the future. Environmental performance dominates the discussion about the industry, both from a public policy and a public relations standpoint. The industry recognizes that its continuing success will depend on a healthy environment. Through the years, the industry has worked hard and invested much to improve and to minimize its impact on the environment, and it has made significant progress.

What drives this progress in environmental stewardship? Government mandates, advances in technology, and better understanding of the impacts of manufacturing processes on the environment have contributed to many of these advances. Also playing a role have been voluntary efforts by the industry.

Voluntary Programs

There are many examples of voluntary industry programs that benefit the environment. One such program, mentioned above, has committed the industry to recovering for recycling and reuse 50 percent of all used paper by the year 2000. Meeting this commitment will ensure that by the year 2000, 40 percent of all fiber used to make pulp and paperboard will be from recovered paper. This figure is up from 25 percent in 1988. Many pulp and paper companies also signed on to the 33/50 program that called for reducing the discharge of 18 toxic substances into the environment by 33 percent by 1992 and by 50 percent by 1995.

EPA's voluntary Energy Star program is designed to reduce emissions of "criteria" and greenhouse gases through increased energy efficiency. It includes Green Lights (focusing on lighting efficiency), Energy Star Computers (focusing on computer and office-system energy efficiency), and Energy Star Buildings (focusing on heating and air conditioning efficiency). These initiatives are directed more at office, warehouse, and light-industrial situations, but forest products companies are looking at the technologies being promoted even if they are not directly participating in the program. One of the latest EPA voluntary initiatives is the Environmental Leadership Program, which recognizes companies that have perfect environmental records.

The Energy Policy Act of 1992 created a voluntary program of reporting of carbon dioxide emissions that is intended to quantify national output of greenhouse gases. In the past, reporting has often been the forerunner of regulation. How this turns out depends largely on the evolution of the science surrounding the greenhouse effect and the international response to the Climate Change Convention of the Earth Summit of 1992.

Government and industry are cooperating increasingly at the highest levels. The best example of this is the President's Council for Sustainable Development, which includes a representative of the forest products industry. The council's activities have not progressed far enough at this writing for their effect on U.S. policy and performance to be assessed.

Voluntary programs are an underutilized means of improving environmental conditions. There is a misconception that voluntary programs do not work because they have no teeth, or that they will not be supported because no one will voluntarily place their company at an economic disadvantage. This way of thinking ignores the fact that corporations are being increasingly open about their environmental practices and are now accountable to the public as well as to regulatory officials. It also flies in the face of the fact that economic benefits can result from good environmental stewardship. Some barriers to successful voluntary programs are mentioned in the last section of this paper.

The Role of Public Opinion and Customer Demands

Less quantifiable but as important as government mandates and industry's voluntary efforts are public opinion and customer demands, whether the result of media coverage or pressure from environmental organizations. Perhaps the most significant of these is news media coverage of environmental news, which has grown dramatically since the first Earth Day in 1970. This coverage has been driven in large part by high-profile environmental disasters such as Three Mile Island, Bhopal, Times Beach, Love Canal, and the *Exxon Valdez*.

The increased media interest in the environment has generated greater interest from the public, and as a result, the pulp and paper industry is being held to ever higher environmental standards. The industry operates in a society that demands that it protect the environment. If it does not, society is going to make it increasingly difficult for the industry to continue to operate.

Because of this realization, pulp and paper companies are taking steps to ensure that they are environmentally responsible and that they clearly communicate their records, progress, and plans. Since 1992, industry management, acting through the industry's association, has adopted a set of environmental, health, and safety principles (American Forest and Paper Association, 1992) and a set of sustainable forestry principles (American Forest and Paper Association, 1995). These principles are a public declaration and commitment by the industry to serve consumer needs for forest products while protecting environmental quality and sustainably managing the forest for present and future generations.

AFPA member companies have promised to promote successful reforestation of nonindustrial private land through cooperative efforts with landowners, federal and state agencies, and other elements of the forestry community. AFPA members will use responsible practices in their own forests and will promote sustainable forestry practices among other forest landowners. Landowners selling timber to AFPA members will be asked to make informed decisions about reforestation.

Many companies in the industry have gone even further to demonstrate environmental stewardship. They have established company-specific principles and codes of environmental conduct and set up departments exclusively focused on ensuring environmental compliance and improving environmental performance beyond regulatory requirements. A number of firms have established a regime of internal facility audits and publish environmental reports covering issues such as emissions data, environmental capital expenditures, and environmental problem areas.

On the resource side, many pulp and paper companies have voluntarily changed their management practices to provide more multiple-use values, such as wildlife habitat and recreational opportunities. In cases where forest land is home to an endangered or threatened species, many industrial landowners have taken the initiative to establish species-protection programs that are also compatible with commercial timber management.

Customer demand is another factor that has changed the way the industry operates. At the vanguard are customer demands for more products with recycled content. The pulp and paper industry also has responded to demands for reduced packaging, and in the United States, customer demand is developing for ECF (elemental chlorine free), TCF, and even unbleached papers.

The ISO has also accepted a mandate to promote changes in corporate culture aimed at benefiting the environment. For example, the ISO Environmental Management Standards subcommittee is developing certain minimum corporate environmental management, training, and communication structures. Companies will need to document their adherence to these to secure ISO certification in the future. AFPA and individual companies are providing input to the development of these standards. Standards for environmental performance will only get more stringent.

In many ways, the easy choices have already been made. In spite of the pulp and paper industry's consistent performance improvements, in the future it will no longer be business as usual.

Near-term issues such as air and water quality, recovery and recycling, and forest practices and timber supply already are giving way to worldwide issues such as global warming, depletion of the ozone layer, and sustainable development. The question the public, the regulators, and the environmental organizations will ask is not, Can the pulp and paper industry create a mill with no emissions? but How soon can they do it? The pressure to accomplish this will be intense. In addition, the industry will be challenged to make even more efficient use of the forest resource, to develop alternative products, and to recover and reuse ever-increasing amounts of its products.

Initiatives by many companies to report environmental information to the public invite public feedback. For this and other reasons, the public will become more and more interested in industrial processes and their effects on the environment and the local community. Also, the public will increasingly have a say about what pulp and paper manufacturers can and cannot do in operating their facilities.

It is important for the industry to meet formally with regulators on a regular basis, perhaps every 2 years, to discuss current environmental problems and how to resolve them. The collaborative efforts between EPA and industry in developing the New Source Performance Standards regulations 20 or so years ago provide a good example. A glossary of environmental terms should be developed as part of this process and agreed upon by regulators, industry, and other stakeholders. For example, "minimum environmental impact" should be defined in terms of adverse effects. It is also important for industry and regulators to work closely together to transfer the technical knowledge needed to develop adequate regulations.

Shaping a Positive Future

The industry has many opportunities to shape a positive future. First, though, the industry needs to make sure its own house is in order—that pulp and paper manufacturers truly are operating in an environmentally responsible way. Companies can improve environmental performance and waste fewer resources through better coordination of their efforts. An example is exporting wastepaper to developing nations so these countries can use their financial resources for things other than building new pulping facilities. Importation of post-consumer recycle material for use in producing paper for packaging, tissue, printing, and so forth might be an effective way to improve the ecological performance of the industry. This would be especially feasible for developing nations with low-cost sources of power.

Implementing more effective environmental management systems can increase environmental awareness among all employees and help ensure company-wide application of environmental principles and policies. It can also ensure that meaningful parameters of performance are communicated to company managers so that corrective action can be taken in a timely manner. In other words, what gets measured gets done.

Too often, the diversity of the forest products industry has led it to follow the lowest-common-denominator strategy: setting goals so that the poorest environmental performers are able to meet them. The industry must set goals that mean something.

In addition, the industry needs to get out in front of the issues and help set policy. The only way to do that is to form strategic partnerships with environmental regulators and environmental organizations to develop collaborative approaches to new environmental requirements. Given the complexity of the environmental issues ahead, the economic consequences of more environmental regulations, and the highly competitive global marketplace, pulp and paper producers do not have the luxury of letting environmental agencies go forward with unilateral command-and-control regulations.

To a large degree, public perception is formed on the basis of aesthetic considerations such as visual effects. Public perception, not necessarily facts, plays a big role in the level of support the industry receives when questions about environmental performance are raised. Aesthetic design concepts used properly in such areas as scrubber plumes, noise abatement, and perimeter fencing would improve the public's impression of what is going on behind company gates.

If the pulp and paper industry is to continue to make meaningful environmental progress, all the players need to be at the table to help establish reasonable environmental goals based on sound scientific principles and to identify more efficient mechanisms to meet those goals. This will not be an easy process. There are strong opinions and plenty of suspicion about motives on all sides. To carry out what the public demands—protection of the environment—and provide what the public needs—a strong economy—the industry must work collaboratively. This spirit of cooperation is one the whole industry has to embrace. No one company can do it alone. Clearly, environmental issues do not respect state, national, or international boundaries. Environmental policies and decisions made in one country or region can affect the environmental debate worldwide. By working together to develop innovative policies and programs, the pulp and paper industry has an opportunity to replace confrontation with cooperation, promote economic growth, and improve environmental quality.

IMPROVING INDUSTRY ENVIRONMENTAL PRACTICES

Setting Environmental Objectives

There is no accepted protocol in the United States for setting long-term environmental objectives. One reason for this, perhaps, is that the history of the command-and-control structure of environmental regulations in this country dis-

THE PULP AND PAPER INDUSTRY	135

	No Cases	One in a Million	Statistically Significant Chance
Any Adverse effect			
Observable adverse effect			
Effect on quality of life			
Morbidity			
Mortality			

FIGURE 4   Matrix of possible human health objectives.

courages cooperative agency-industry development of environmental goals. This is a major barrier to progress.

The first step is to decide what overall objectives the country should be striving for. For each major environmental issue, such as human health effects, an objective should be picked using a scientifically based public consensus process. Figure 4 illustrates the range of possible objectives.

Once made, the selection should be adhered to. Oftentimes, Congress and the federal agencies seem to decide what is important based on current media and public attention or political payoff. These forces typically exert an effect for a relatively short period, typically about 3 years. This is much less time than it takes for environmental problems to respond to corrective action, often 15 to 20 years. What happens can be compared to an automobile cruise control. Like environmental rules, a cruise control is intended to establish, and minimize deviation from, a desired performance. If the throttle (Congress) reacts too quickly to changes in speed (environmental results), then speed varies wildly, the objective of reduced deviation is not met, and fuel is wasted.

The short time horizon for "throttle" adjustments also causes problems related to the capital-intensive nature of the industry (indeed, of any heavy industry). First, the industry cannot afford to change manufacturing processes to incorporate new technology every few years. Modernization of facilities tends to reduce the amount of pollutants generated and released. A regulatory time scale that allows time for modernization can benefit the business as well as the environment, because modernized facilities will also be more efficient, produce higher-quality products, and be more profitable. Second, the capital-intensive nature of the industry does not allow for sufficient funds to advance the research and development of environmental technology. Therefore, industry should supplement its own research effort by collaborating with federal research initiatives.

The next step in setting goals is to decide which environmental problems are worth tackling, keeping in mind the objective agreed to earlier. There is tremendous variation today in the benefit society receives per environmental dollar, depending on where it is spent. This has long been recognized in industry and is now being discussed within U.S. agencies. Congress and the regulators seem too slow in appreciating that the resources for dealing with any issue are limited and that priorities must be set for environmental efforts according to environmental damage avoided per unit cost. For example, as measurement capabilities continue to improve, it might become counterproductive from an environmental standpoint to reduce releases below the point of adequate protection. That is, the mere presence of a measurable amount of a substance should not warrant regulatory control. A scientific consensus process should be developed to place environmental issues on a priority list as they arise. The ability to address items on the list must match available money.

Many problems would be eliminated if objectives and goals were set through a public process. For example:

- There would be greater agreement among all interested parties that the costs of environmental management were worthwhile; hence, there would be a higher level of compliance and fewer legal and administrative costs.
- There would be less-frequent shifting of emphasis among environmental areas and less change within bodies of regulations. The benefits of this would include less external oversight of and better capital planning within companies. Resources that have been used to keep track of regulatory change could be diverted into economically and environmentally beneficial activities. Also, there might be fewer lawsuits.
- Regulators could devote more of their resources to improving regulatory efficiency by streamlining existing regulations and by analyzing and understanding the connectedness of regulations.

Enhancing Public Understanding of Environmental Science

Better public appreciation for the impracticality—often impossibility—of achieving zero emissions or completely eliminating potentially harmful substances is needed. Careless dissemination of environmental information can lead to problems in a society in which response to environmental needs is developed democratically. Here is a partial list of steps that can be taken to improve the public's understanding of environmental science:

- Every new law or regulation mandating greater public access to information (e.g., toxic-release inventories) should provide for public education. If lawmakers are going to require companies to publish technical information, they must provide the public with the tools to understand it

and not leave the task to companies ill equipped to handle it. The public (including the legislators themselves) needs and deserves to understand this information and the issues. Public misunderstanding leads to bad environmental law. Environmental rules exist, after all, to provide greater security, not to cause panic. The movement in the United States toward using risk-assessment techniques to help set priorities for environmental action could be a vehicle to correcting misconceptions of the past.
- By being open and candid, companies can increase their own credibility, that of their industry, and that of the private sector as a whole. If industry takes this stance, any concerns it brings to the public debate are more apt to be heard and heeded.
- The media must learn or be made aware of the stakes involved in environmental issues. Sensationalism on this topic can result in the waste of considerable resources.
- A process needs to be developed that involves all stakeholders in a discussion of topics such as sustainable development and biodiversity. Stakeholders also need to be informed about scientific research on such questions as the ecological functions of trees and the point when new growth is considered old growth.

Creating Incentives and Encouraging Flexibility

Voluntary programs that recognize achievement and provide companies financial returns have been successful at creating incentives for constructive change. Examples of incentives include less-stringent monitoring programs (as now exist in occupational safety rules) and "banking" of voluntarily reduced emissions, or the receipt of tax advantages for such reductions.

Some existing voluntary programs that could provide economic benefits to participants are hampered by being too rigid. For example, EPA's Green Lights program will not allow a company to sign up only its lighting-intensive facilities. By being all or nothing, the program has probably delayed some reductions in air emissions that might otherwise have been achieved. Similarly, the EPA's Environmental Leadership Program has only one level of recognition—a perfect compliance record. It is easier for light industries to meet this standard than heavy industries such as pulp and paper. Voluntary programs should respect differences among the various industry sectors and should not be allowed to become command-and-control regulations in the future. For example, the 1993 federal paper-recycling rules (Executive Order 12873) have challenged the industry's preexisting recycling efforts.

Voluntary effort recommended by the industry in the 1988 Wetlands Policy Forum deserves special mention here as an opportunity for realizing environmental and economic benefits in forest management. The effort provides for beneficial use of forested wetlands under a voluntary program of habitat protection in lieu of outright taking under the existing wetlands policy. Such taking not only infringes on property rights and values but also is ecologically counterproductive because it removes any economic incentive to maintain such lands in forest cover.

The industry has made some related suggestions regarding the protection of endangered species. Here again, a means of compensating landowners for the effective taking of their property for the common good of protection of species must be developed. Another flaw in the current system is that it does not allow for adequate scientific peer review of decisions surrounding the identification and protection of species. In this respect, it differs from most other regulatory scientific protocols. The application of sound science, in addition to public participation, is an important component for regulatory policies, including those for protection of endangered species. In addition, effective environmental policy by the government requires cost-benefit analyses and consideration of environmental trade-offs. For example, in the paper industry, the trade-offs involved in recycling, which would cause increased reliance on external energy sources (e.g., fossil fuels), must be taken into account.

Facility Modernization

Another barrier to environmental improvement in the industry has to do with land and groundwater use, zoning, and difficulties in obtaining permits. Modernization of existing facilities sometimes presents logistical problems that are difficult to overcome, yet land-use competition and the perception that industrial facilities pose hazards to surrounding residents discourage the siting of facilities on new grounds, which are generally cleaner and more efficient. Here again, better communication to the public by EPA and industry representatives about the good science behind risk assessment is important.

Creating a Climate for Innovation

The recent explosion in environmental enforcement in the United States is stifling innovation because the penalty if an innovation fails to meet requirements is too great. A broadening of statutory and regulatory exemptions and tax incentives for innovative environmental technology will have a long-term beneficial effect on the environment.

Encouraging Cautious Consideration of Life-Cycle Assessment

How are life-cycle assessment initiatives handling forestry issues? The state of the art in setting boundaries for the life-cycle inventory phase seems to focus on resource depletion and does not yet include other proper elements of the forest life cycle, such as sustainability. Forestry issues are politically hot, so it is difficult to steer LCA discussions away from this one resource and toward the broader issue of extraction of resources in general. Obviously, each use of a natural resource affects streams, biodiversity, and so forth in its own way. In recent life-cycle-based methods, from LCA inventory studies to national environmental labeling schemes, the approach in forestry has been to provide evidence of sustainable regional or national yields (growth greater than or equal to harvest), because the claim of renewability of a resource can be supported only by accounting across a controlled time and land area. Johnston (1997) has reviewed approaches, sources, and uncertainties associated with LCA in the context of the pulp and paper industry.

LCA activity is also part of the ISO standards initiative commissioned in 1991 by ISO Technical Committee 207. The intent is to establish some degree of conformity in LCA methodology, as well as in environmental management systems, environmental auditing, ecolabeling, environmental performance evaluation, and incorporation of environmental aspects in other product standards.

Recycling initiatives already have had an important impact on the development of LCA. Efforts to establish a formal hierarchy and preferred methods have tended to interfere with otherwise more scientific and technical efforts. Such biases have been seen in drafts of the ecolabel criteria of the European Community, as well as in other practitioners' approaches to the implementation of LCA methodology, and have the potential to damage the credibility of LCA. In future LCA developmental activities, representatives of the forest products industry must be alert to two key sources of bias: the mechanics of allocating environmental burdens between recycled and virgin materials, and prejudices about acceptable methods of waste disposal.

In the case of disposal options, the development of a structural hierarchy could blur the objective evaluation of solid-waste management approaches. Solutions that are better implemented for reasons of timing or location could be discarded accidentally. New analyses have been emerging showing techniques and circumstances in which disposal for energy recovery can be better for the environment than after-market recycling.

LCA studies are designed with pre-set boundaries. This feature is necessary to manage the required logistical and scientific data gathering and analysis. In the case of the forest products industry, it is important in any LCA to consider how the industry manages its forest resources and the relationship of that management to a particular line of forest products. Because people often react emotionally to the loss of wooded areas, articulating scientific positions is difficult and is a burden other sectors, such as agriculture, do not have to bear. A proper analogy in this case is to an agricultural crop such as corn: no one thinks of cereal production as destroying corn plants.

It is well to continue efforts to show that silviculture has many parallels with agriculture and that other resource uses have adverse impacts, too, but these efforts will not be sufficient to put discussions about the forest resource back on a scientific track. At least one other essential step is to maintain a good track record on silvicultural practices and to make these practices widely known.

Forestry practices should not create accountability problems in the inventory phase of an LCA. The evaluation process of LCA should encompass credits for renewable forestry activities and the transformation of nonproductive farmland into productive forestland. However, the state of the art in the impact phase of LCA is not sufficiently developed in silviculture and forest management practices to permit, at this writing, an accurate prediction of the potential future effect of a particular practice on LCA results.

There are two energy-related considerations that might further improve the LCA process for forest products. One is to emphasize, as stated before, that more than 50 percent of the energy needs of the industry are satisfied at present by renewable biomass energy. This energy supply is carbon neutral with regard to potential climate change. Incremental energy demands in the industry over and above this are met mostly by fossil fuel. The other consideration is to make use of the energy in recovered paper, when either the amount of recovered paper exceeds, or its quality is less than, the optimal for recycling purposes. Such utilization will be limited in many cases by equipment costs and location. Nevertheless, it is a power-generation alternative, which in terms of LCA could be justified in certain locations and at certain times. A recent example is the change in direction in the German packaging regulations, providing for less recycling and more conversion of waste to energy, in response to hard political and economic realities in that country.

Finally, it is important for the industry to be involved in the international development of the LCA methodology. Once international consensus is achieved, it will be difficult to change. The consideration of life-cycle aspects in industrial operations and products can lead to improved practice. However, competing interests from various industrial sectors (e.g., paper and plastic) and the associated economic stakes create a difficult environment for developing scientifically sound, objective criteria for evaluating environmental impacts of manufacturing processes and appropriate regulatory responses. The many questions raised about LCA and the inherent biases in the technique make such assessments inappropriate for regulatory purposes. They are best used for internal decision making in companies.

REFERENCES

American Forest and Paper Association (AFPA). 1992. Environmental, Health and Safety Principles. Washington, D.C.: AFPA.

American Forest and Paper Association (AFPA). 1994. Pollution Prevention Report. Washington, D.C.: AFPA.

American Forest and Paper Association (AFPA). 1995. Sustainable Forestry Implementation Guidelines. Washington, D.C.: AFPA. Also available on the Internet, <http://www.afandpa.org/forestry/guidelines.html>.

Balzhiser, R.E. 1989. Meeting the near-term challenge for power plants. Pp 95–113 in Technology and Environment, J.H. Ausubel and H.E. Sladovich, eds. Washington, D.C.: National Academy Press.

Canadian Forest Service (CFS). 1993. Selected Forest Statistics Canada 1992. Ontario, Canada: CFS.

Casey, J.P. 1983. Pulp and paper. Pp. 63–79. Chemistry and Chemical Technology, 3rd ed. New York, N.Y.: John Wiley and Sons.

Dence, C.W., and D.W. Reeve. 1996. Pulp Bleaching: Principles and Practice. Atlanta, Ga.: Technical Association of the Pulp and Paper Industry.

Folke, J. 1994. Environmental effects from modern bleach plant manufacturing. Paper presented at the National Academy of Engineering International Conference on Industrial Ecology, Irvine, Calif., May 9–11.

Johnston, R. 1997. A critique of life-cycle analysis: Paper products. Pp. 225–233 in The Industrial Green Game: Implications for Environmental Design and Management, D.J. Richards, ed. Washington, D.C.: National Academy Press.

Miller Freeman (MF). 1994. Lockwood-Post's Directory of the Paper and Allied Trades. 1994. San Francisco: MF.

Saltman, D. 1983. Pulp and Paper Primer. Atlanta, Ga.: Technical Association of the Pulp and Paper Industry.

Technical Association of the Pulp and Paper Industry (TAPPI). 1992. TAPPI Pulping Conference Proceedings. Atlanta, Ga.: TAPPI.

Biographical Data

A. DOUGLAS ARMSTRONG is retired manager for pulp and paper feasibility at Georgia-Pacific Corp., where he served in various plant and management positions for over 40 years addressing operational and environmental concerns.

PATRICK R. ATKINS is director of environmental control at Alcoa. He joined Alcoa in Pittsburgh in 1972, after 4 years as a professor in environmental health engineering at the University of Texas at Austin, where he taught engineering, industrial hygiene, and ecology and directed M.S. and Ph.D. research projects for 23 students. In 1973, Atkins became Alcoa's manager of environmental control. He was named to his present position in 1980. Atkins also served as the company's chief environmental engineer from 1982 to 1984. Author of over 50 technical articles and editor of 2 books, he is a member of the American Society of Civil Engineers, the Water Pollution Control Federation, the National Society of Professional Engineers, and the Engineering Society of Western Pennsylvania. Atkins represents Alcoa on the environmental committee of the International Primary Aluminum Institute, the Business Roundtable, National Association of Manufacturers, and other national and international groups. He is a registered professional engineer in the states of Texas and Pennsylvania and serves as an adjunct professor at the University of Pittsburgh, teaching industrial waste-treatment technology. Atkins has a B.S. in civil engineering from the University of Kentucky and an M.S. and Ph.D. in environmental engineering from Stanford University.

KEITH M. BENTLEY is director of environmental engineering technical support at Georgia-Pacific Corp. In that capacity, he manages a group of engineers that provides technical expertise to Georgia-Pacific's plants and mills. Bentley has experience reviewing plant activities to ensure compliance with applicable regulations; assisting in the design and selection of environmental control equipment; developing environmental policy; conducting regulatory development and analysis; and negotiating emission permits, variances, and compliance schedules with various regulatory agencies. He has over 23 years of experience in environmental engineering, the last 17 with Georgia-Pacific. Bentley has also held leadership positions on various committees within the Technical Association of the Pulp and Paper Industry and the American Forest and Paper Association. He has a B.S. in chemical engineering from the University of South Carolina.

CHARLES G. CARSON III is vice president of environmental affairs at U.S. Steel. His responsibilities include overseeing U.S. Steel's environmental compliance and improvement activities and coordinating the company's relations with various environmental agencies and groups. Carson joined U.S. Steel in 1970 as a research engineer and progressed through a series of technical and management positions in various research and development areas. In 1985, he moved into U.S. Steel's commercial operations as manager of product development in the Tin Mill Products department. Carson was promoted in 1990 to general manager of Tin Mill Products. He has a B.S. in chemistry from Williams College and an M.S. and Ph.D. in metallurgy from The Pennsylvania State University.

PRESTON S. CHIARO is vice president of technical services at Kennecott Corp. He is primarily responsible for Kennecott's compliance with all federal, state, and local environmental requirements for all of its active mining and mineral processing operations as well as for its exploration, acquisition, and reclamation activities. Chiaro also oversees environmental audits and due-diligence assessments, provides assistance for getting permits, helps to promote industry concerns among regulators, and promotes energy efficiency and waste-minimization awareness. He joined Kennecott in 1991 to help direct a massive cleanup of historic mine wastes at Kennecott Utah Copper near Salt Lake City. In October 1992, Chiaro was named Kennecott's vice president of environmental affairs. Prior to joining Kennecott, he managed a large environmental cleanup contract for Ebasco Environmental. Chiaro has a B.S. and M.S. (cum laude) in environmental engineering from Rensselaer Polytechnic Institute. He is a registered professional engineer in five states.

ROBERT A. FROSCH is a senior research fellow at Center for Science and International Affairs of the John F. Kennedy School of Government at Harvard University and senior fellow at the National Academy of Engineering. In 1989, he revived, redefined, and popularized the term industrial ecology, and his research has focused on this field in recent years, especially in metals-handling industries. In 1993, Frosch retired as vice president of General Motors Corp., where he was in charge of the North American Operations Research and Development Center. After doing research in underwater sound and ocean acoustics, he served for a dozen years in a number of government positions, including deputy director of Advanced Research Projects Agency of the Department of Defense, assistant secretary of the Navy for research and development, assistant executive director of the United Nations Environment Program, and administrator of the National Aeronautics and Space Administration. Frosch is a member of the National Academy of Engineering and holds a Ph.D. in theoretical physics from Columbia University.

ANN B. FULLERTON is president of The Fullerton Group, a media relations consulting firm. She has more than 10 years of product and program marketing experience for Fortune 500 companies, industry associations, and government programs. Fullerton's expertise includes design implementation and evaluation of marketing programs with substantive experience in the areas of high technology, science, and the environment. She has served as marketing advisor to senior management at AT&T, Northern Telecom, and Digital Equipment Corp. and has been recognized for her organizational and leadership abilities in taking ideas from concept to implementation. Fullerton has also served on the board of directors and chaired communications subcommittees of several industry associations, including the American Electronics Association, the Computer Business Equipment Manufacturers, the Center for Office Technology, and the Industry Cooperative for Ozone Layer Protection. She holds a B.S. (cum laude) in public relations from Boston University and an M.S. in communications management from Simmons College.

SERGIO F. GALEANO is manager of environmental programs for Georgia-Pacific Corp. His responsibilities are in product safety and assurance; policy and technology issues in connection with the environmental attributes of products; and trade and competitiveness issues. Galeano also chairs various committees of domestic and international forest products industry groups. He has received a number of honorary awards, among them membership in Phi Kappa Phi and the 1995 TAPPI Environmental Division Technical Award for outstanding technical contributions in the area of environmental protection. Galeano holds more than a dozen patents and has published extensively. He has an M.S. in civil engineering from the University of Havana and an M.S.E. and Ph.D. in environmental science and engineering, respectively, from the University of Florida at Gainesville. Galeano is a registered professional engineer and a diplomat of the American Academy of Environmental Engineers.

THOMAS E. GRAEDEL is professor of industrial ecology at the Yale School of Forestry and Environmental Studies, a position he assumed after 27 years as Distinguished Member of Technical Staff at AT&T Bell Laboratories. He was the first atmospheric chemist to study the atmospheric reactions of sulfur and the concentration trends in methane and carbon monoxide. As a corrosion scientist, Graedel devised the first computer model to simulate the atmospheric corrosion of metals. This work led to a voluntary position as consultant to the Statue of Liberty Restoration Project in 1984–1986. One of the founders of the emerging discipline of industrial ecology, he co-authored the first textbook in the field and has lectured widely on its implementation and implications. Graedel has published 9 books and more than 200 scientific papers. He holds a B.S. from Washington University, an M.A. from Kent State University, and an M.S. and Ph.D. from the University of Michigan.

G. FRANK JOKLIK began his career as an exploration geologist with Kennecott Corp. in 1954, heading up minerals projects in Canada and the United States. After 10 years, he joined Amax to manage the development of the Mt. Newman iron ore mines, Western Australia, and other projects. Joklik was subsequently elected a corporate vice president. In 1974, he resumed his career with Kennecott and, after several promotions, became president in 1980. He continued in that role through several changes of ownership. He retired from Kennecott in June 1993 and now operates from his office in Salt Lake City. A highlight of his career was the revitalization of Bingham Canyon mine where, through cost reduction and investment in the replacement of antiquated plant, a high-cost, inefficient operation was converted into one of the world's lowest-cost, most-productive, and environmentally clean producers of copper and precious metals. Under his management, Kennecott developed five new precious and base-metal mines in the United States, discovered and defined the giant Lihir gold deposit in Papua New Guinea, and through acquisition became the leading coal producer in the Powder River Basin of Wyoming and Montana. Joklik is a member of the National Academy of Engineering and a distinguished member of the Society for Mining, Metallurgy, and Exploration. He was elected Copper Club Man of the Year for 1988 and, in 1991, received the AIME William Lawrence Saunders Gold Medal for distinguished service to the mining industry. Joklik was born in Vienna, Austria, and grew up in Australia. After schooling at Cranbrook, he attended the University of Sydney, where he received B.Sc. (First Class Honors) and Ph.D. degrees in geology. In 1953, he came to the United States as a Fulbright Scholar at Columbia University.

ROBERT A. LAUDISE is adjunct chemical director at AT&T Bell Laboratories, where he is responsible for chemistry-related R&D throughout AT&T Bell Labs, adjunct professor of materials science at MIT, and adjunct professor of ceramics at Rutgers University. He has a special interest in green materials and processes. Laudise joined AT&T Bell Laboratories in 1956 and has served as director of materials research, physical and inorganic chemistry research, and materials and processing. His interests include solid state chemistry, materials science and conservation, and crystal growth. Laudise has spent considerable time studying hydrothermal crystallization and the preparation of piezoelectric, laser, nonlinear optical, and related materials. Most commercial processes for preparing crystalline quartz are based on his work. Laudise is the author of the book *The Growth of Single Crystals* and more than 150 publications on crystal growth and related fields. He holds 12 patents and has received numerous prizes and awards, including the A.D. Little Fellowship and the 1976 Sawyer Prize. Laudise is the past president of the International Organization of Crystal Growth, is a member of the American Chemical Society, National Academy of Engineering, National Academy of Sciences, and is a fellow of the American Association for the Advancement of Science. He chairs the National Materials Advisory Board and is editor-in-chief of the *Journal of Materials Research*. Laudise has a B.S. in chemistry from Union College and a Ph.D. in inorganic chemistry from MIT.

KENNETH J. MARTCHEK is program manager of pollution prevention and life-cycle analysis for Alcoa. He reports to the corporate director of environmental affairs and is responsible for awareness, development, and promotion of waste-reduction and product-stewardship initiatives for Alcoa worldwide. Martchek joined Alcoa in 1979 after 3 years with U.S. Research Laboratories, where he was responsible for process simulations of ore processing and blast-furnace operations. As a senior project engineer at Alcoa Technical Center, he supervised the operation of pilot plant facilities in the production of high-purity aluminum and solar-grade silicon. Martchek joined the environmental engineering staff of Alcoa in 1983 and directed a number of commercial project installations in water and waste treatment. In 1992, he received Alcoa's Davis Award for the R&D, design, and successful startup of a biological treatment facility associated with a new, high-speed electrocoating operation. Prior to assuming his present position, Martchek served as engineering manager for Alcoa's effort to develop ceramic electronic packaging. He is a member of the American Institute of Chemical Engineers and American Society for Quality. Martchek has a B.S. in chemical engineering from the University of Pittsburgh and an M.S. in chemical and biochemical engineering from Rutgers University. He is a licensed professional engineer in the state of Pennsylvania.

ELIZABETH H. MIKOLS is manager of environmental affairs for CBR-HCI Construction Materials Corp., one of the largest producers of portland cement in the United States and Canada. Prior to coming to CBR-HCI in 1989, she held environmental positions at Lonestar Industries, another major cement, concrete, and construction materials company. In addition to her 13 years in the cement industry, Mikols had 9 years of experience working in environmental regulatory agencies in New York and Connecticut. She participates on several environmental committees and task groups for industry associations and has delivered presentations at technical conferences and public meetings on a variety of environmental subjects connected with the cement industry. Mikols is a member of the Air & Waste Management Association, a nationwide association of environmental professionals, and serves on the Dean's Strategic Advisory Council for the Yale School of Forestry and Environmental Studies. She has a bachelor's degree from the University of California at Berkeley and a master's degree from the School of Forestry and Environmental Studies at Yale University.

ROBERT J. OLSZEWSKI is director of environmental affairs, forest resources, at Georgia-Pacific Corp. Prior to joining industry, he worked for 6 years as Florida's forest hydrologist. In that position, Olszewski was responsible for implementing the forestry non-point-source pollution element of the state's water-quality plan and actively trained Florida's forestry community in silvicultural best-management-practice applications. His next job was with the Florida Forestry Association, where his duties included working with the forestry community, environmental organizations, the Florida legislature, and federal, state, and local governments on a variety of environmental and land-use regulations affecting forestry operations in the state. In March 1993, Olszewski accepted a newly created position with Georgia-Pacific, working for the Forest Resources Business Unit and the Environmental Policy, Training, and Regulatory Group on environmental issues affecting forestry operations at the company. He assumed his current post in January 1996. Olszewski is active with the Society of American Foresters (SAF), having served as Florida section chairman and as chairman of Georgia SAF's Chattahoochee chapter, and is chairman of the American Forest and Paper Association's Forest Wetland and Nonpoint Source Committee and Global Forestry Committee. He has a B.S. in forestry from Michigan Technological University and an M.S. in forest hydrology from the University of Georgia, Athens.

DEANNA J. RICHARDS is associate director of the National Academy of Engineering's Program Office and also directs the Academy's program on Technology and Environment (T&E). Hired in 1991 to launch the Academy's environmental effort, she has led groundbreaking work that has helped establish the field of industrial ecology. The T&E program has focused on technological trajectories of large-scale systems, as well as on best practices in environmental design and management in the manufacturing and service industries. Richards has directed several projects related to industry efforts to integrate environmental considerations in decision making. Before joining the NAE, Richards was an assistant professor of environmental engineering and worked for several years in that field. She has a B.S. (honors) in civil engineering from the University of Edinburgh and M.S. and Ph.D. degrees, also in civil engineering, from the University of Pennsylvania.

GAIL A. SMITH manages corporate environmental communications for Georgia-Pacific Corp., where she has worked since 1984. She is active in industry groups, serving on communications committees for the American Forest and Paper Association and the Great Lakes Water Quality Coalition. Smith is also a member of the International Association of Business Communicators. She is a graduate of Brenau College in Gainesville, Ga., where she majored in business administration and English.

JONATHAN R. SMITH JR. is president of Rigdon Engineering, a consumer products engineering and development firm. His career in the pulp and paper industry began in 1969, when he became research engineer at Chesapeake Corp. of Virginia. Smith later joined the Mead Corp. as senior consultant, corporate human and environmental protection department. He then held increasingly responsible positions at Georgia-Pacific Corp., becoming senior manager of environmental affairs in 1991. Smith has published numerous articles in peer-reviewed journals and holds four U.S. patents. He is a diplomate of the American Academy of Environmental Engineers, a member of the Technical Association of the Pulp and Paper Industry, American Institute of Chemical Engineers, and Brunswick-Golden Isles Chamber of Commerce. Smith has a B.S. in chemical engineering from the University of Virginia and did graduate work in systems engineering at Ohio University and in water-pollution control at Vanderbilt University.

IAN M. TORRENS is head of power generation for Shell International in London, where he is responsible for business development activities involving use of Shell gas and coal fuels by electric utilities and independent electric power developers worldwide. He also represents Shell International Gas and Shell Coal International in a number of international bodies such as the International Energy Agency and the World Coal Institute. Prior to assuming his current position in 1995, Torrens was director of the Environment Control Business Unit at the Electric Power Research Institute (EPRI). At EPRI, he directed R&D in several areas relating to electric power generation: SO_2, NO_x, and particulate control systems; CO_2 mitigation; and waste and water management, including management of potentially toxic substances in air and water. Between 1973 and 1987, he was at the Organization for Economic Cooperation and Development (OECD) in Paris, where he worked on energy issues in the OECD's International Environment Directorate. The author of three books and several papers, Torrens has a B.Sc. (first class honours) in physics from Queen's University, Belfast, and a Ph.D. in nuclear physics from the University of Cambridge.

KURT E. YEAGER is president and chief executive officer of the Electric Power Research Institute (EPRI). After joining EPRI in 1974, he held a series of R&D management positions before being named, in 1990, as senior vice president of technical operations, responsible for the integrated management of all EPRI technical programs. In 1994, Yeager became senior vice president for strategic development and, in 1995, executive vice president and chief operating officer. Before joining EPRI, he was director of energy R&D planning for the Environmental

Protection Agency's Office of Research and associate head of enviornmental systems development at the MITRE Corp. Yeager has a bachelor's degree from Kenyon College and completed postgraduate studies in chemistry and physics at Ohio State and the University of California at Davis. He is a distinguished graduate of the Air Force Nuclear Research Officers Program and a fellow of the American Society of Mechanical Engineers.